COMPUTER-AIDED DESIGN AND VLSI DEVICE DEVELOPMENT
SECOND EDITION

THE KLUWER INTERNATIONAL SERIES IN ENGINEERING AND COMPUTER SCIENCE

VLSI, COMPUTER ARCHITECTURE AND DIGITAL SIGNAL PROCESSING

Consulting Editor

Jonathan Allen

Other books in the series:

Logic Minimization Algorithms for VLSI Synthesis, R. K. Brayton, G. D. Hachtel, C. T. McMullen, and A. L. Sangiovanni-Vincentelli. ISBN 0-89838-164-9.
Adaptive Filters: Structures, Algorithms, and Applications, M.L. Honig and D. G. Messerschmitt. ISBN 0-89838-163-0.
Introduction to VLSI Silicon Devices: Physics, Technology and Characterization, B. El-Kareh and R. J. Bombard. ISBN 0-89838-210-6.
Latchup in CMOS Technology: The Problem and its Cure, R. R. Troutman. ISBN 0-89838-215-7.
Digital CMOS Circuit Design, M. Annaratone. ISBN 0-89838-224-6.
The Bounding Approach to VLSI Circuit Simulation, C. A. Zukowski. ISBN 0-89838-176-2.
Multi-Level Simulation for VLSI Design, D. D. Hill, D. R. Coelho ISBN 0-89838-184-3.
Relaxation Techniques for the Simulation of VLSI Circuits, J. White and A. Sangiovanni-Vincentelli ISBN 0-89838-186-X.
VLSI CAD Tools and Applications, W. Fichtner and M. Morf, Editors ISBN 0-89838-193-2
A VLSI Architecture for Concurrent Data Structures, W. J. Dally ISBN 0-89838-235-1.
Yield Simulation for Integrated Circuits, D. M. H. Walker ISBN 0-89838-244-0.
VLSI Specification, Verification and Synthesis, G. Birtwistle and P. A. Subrahmanyam ISBN 0-89838-246-2.
Fundamentals of Computer-Aided Circuit Simulation, W. J. McCalla ISBN 0-89838-248-3.
Serial Data Computation, S. G. Smith and P. B. Denyer ISBN 0-89838-253-X.
Phonologic Parsing in Speech Recognition, K. W. Church ISBN 0-89838-250-5.
Simulated Annealing for VLSI Design, D. F. Wong, H. W. Leong, C. L. Liu ISBN 0-89838-256-4.
Polycrystalline Silicon for Integrated Circuit Applications, T. Kamins ISBN 0-89838-259-9.
Fet Modeling for Circuit Simulation, D. Divekar ISBN 0-89838-264-5.
VLSI Placement and Global Routing Using Simulated Annealing, Carl Sechen ISBN 0-89838-281-5.
Adaptive Filters and Equalisers, Bernard Mulgrew and Colin F. N. Cowan ISBN 0-89838-285-8.

COMPUTER–AIDED DESIGN AND VLSI DEVICE DEVELOPMENT
SECOND EDITION

KIT MAN CHAM
SOO-YOUNG OH
JOHN L. MOLL
KEUNMYUNG LEE
PAUL VANDE VOORDE
Hewlett-Packard Laboratories

DAEJE CHIN
Samsung Semiconductor

KLUWER ACADEMIC PUBLISHERS
Boston/Dordrecht/London

Distributors for the United States and Canada
Kluwer Academic Publishers
101 Philip Drive
Assinippi Park
Norwell, Massachusetts 02061, USA

Distributors for the UK and Ireland:
Kluwer Academic Publishers
Falcon House, Queen Square
Lancaster LA1 1RN, UNITED KINGDOM

Distributors for all other countries:
Kluwer Academic Publishers Group
Distribution Center
Post Office Box 322
3300 AH Dordrecht, THE NETHERLANDS

Library of Congress Cataloging-in-Publication Data

Computer-aided design and VLSI device development / Kit Man Cham ... [et al.]. — 2nd ed.
   p. cm. — (The Kluwer international series in engineering and computer science: 53. VLSI, computer architecture, and digital signal processing)
   Includes bibliographies and index.
   ISBN 0-89838-277-7
   1. Integrated circuits—Very large scale integration—Design and construction—Data processing. 2. Computer-aided design. I. Cham, Kit Man. II. Series: Kluwer international series in engineering and computer science ; SECS 53. III. Series: Kluwer international series in engineering and computer science. VLSI, computer architecuture and digital signal processing.
TK7874.CG474   1988
621.395—dc19   88-23464
                                                                                                          CIP

Copyright © 1988 by Kluwer Academic Publishers, Boston.

All rights reserved. No part of this publication may be reproduced, stored in a retrieval system, or transmitted in any form or by any means mechanical, photocopying, recording, or otherwise, without the prior written permission of the publishers, Kluwer Academic Publishers, 101 Philip Drive, Assinippi Park, Norwell, Massachusetts 02061.

Printed in the United States

CONTENTS

PREFACE

OVERVIEW   1

PART A : NUMERICAL SIMULATION SYSTEMS   11

Chapter 1. Numerical Simulation Systems   13
  1.1 History of Numerical Simulation Systems   13
  1.2 Implementation of a Numerical Simulation System   15

Chapter 2. Process Simulation   23
  2.1 Introduction   23
  2.2 SUPREM: 1-D Process Simulator   25
  2.3 SUPRA : 2-D Process Simulator   46
  2.4 SOAP : 2-D Oxidation Simulator   58

Chapter 3. Device Simulation   71
  3.1 GEMINI : 2-D Poisson Solver   71
  3.2 CADDET : 2-D 1-Carrier Device Simulator   88
  3.3 PISCES-II : General-Shape 2-D 2-Carrier Device Simulator   102

Chapter 4. Parasitic Elements Simulation   129
  4.1 Introduction   129
  4.2 SCAP2 : Two-Dimensional Poisson Equation Solver   130
  4.3 FCAP3 : Three-Dimensional Poisson Equation Solver   136

PART B : APPLICATIONS AND CASE STUDIES   141

Chapter 5. Methodology in Computer-Aided Design for Process and Device Development   143
  5.1 Methodologies in Device Simulations   143
  5.2 Outline of the case studies   149

Chapter 6. SUPREM III Application   151
  6.1 Introduction   151

6.2 Boron Implant Profiles — 153
6.3 Thin Oxide Growth — 154
6.4 Oxygen Enhanced Diffusion of Boron — 157
6.5 Shallow Junctions — 163

Chapter 7. Simulation Techniques for Advanced Device Development — 167

7.1 Device Physics for Process Development — 167
7.2 CAD Tools for Simulation of Device Parameters — 175
7.3 Methods of Generating Basic Device Parameters — 180
7.4 Relationship between Device Characteristics and Process Parameters — 187

Chapter 8. Drain-Induced Barrier Lowering in Short Channel Transistors — 197

Chapter 9. A Study of LDD Device Structure Using 2-D Simulations — 211

9.1 High Electric Field Problem in Submicron MOS Devices — 211
9.2 LDD Device Study — 213
9.3 Summary — 229

Chapter 10. The Surface Inversion Problem in Trench Isolated CMOS — 233

10.1 Introduction to Trench Isolation in CMOS — 233
10.2 Simulation Techniques — 236
10.3 Analysis of the Inversion Problem — 238
10.4 Summary of Simulation Results — 242
10.5 Experimental Results — 246
10.6 Summary — 248

Chapter 11. Development of Isolation Structures for Applications in VLSI — 251

11.1 Introduction to Isolation Structures — 251
11.2 Local Oxidation of Silicon (LOCOS) — 253
11.3 Modified LOCOS — 259
11.4 Side Wall Masked Isolation (SWAMI) — 261
11.5 Summary — 267

Chapter 12. Transistor Design for Submicron CMOS Technology — 271

12.1 Introduction to Submicron CMOS Technology — 271

Contents

12.2 Development of the Submicron P-Channel MOSFET Using Simulations ... 276
12.3 N-Channel Transistor Simulations ... 293
12.4 Summary ... 298

Chapter 13. A Systematic Study of Transistor Design Trade-offs ... 301

13.1 Introduction ... 301
13.2 P-Channel MOSFET with N^- Pockets ... 302
13.3 The Sensitivity Matrix ... 306
13.4 Conclusions ... 313

Chapter 14. MOSFET Scaling by CADDET ... 315

14.1 Introduction ... 315
14.2 Scaling of an Enhancement Mode MOSFET ... 317
14.3 Scaling of a Depletion Mode MOSFET ... 326
14.4 Conclusions ... 333

Chapter 15. Examples of Parasitic Elements Simulation ... 335

15.1 Introduction ... 335
15.2 Two-Dimensional Parasitic Components Extraction ... 335
15.3 Three-Dimensional Parasitic Components Extraction ... 348

APPENDIX ... 359
Source Information of 2-D Programs

TABLE OF SYMBOLS ... 361

SUBJECT INDEX ... 369

ABOUT THE AUTHORS ... 377

Preface to Second Edition

This second edition incorporates substantial enhancements to the first editon.

This book is concerned with the use of Computer-Aided Design (CAD) in the device and process development of Very Large Scale Integrated Circuits (VLSI). The emphasis is in Metal-Oxide-Semiconductor (MOS) technology. State-of-the-art device and process development are presented.

This book is intended as a reference for engineers involved in VLSI development who have to solve many device and process problems. CAD specialist will also find this book useful since it discusses the organization of simulation system, and also presents many case studies where the user applies the CAD tools in different situations. This book is also intended as a text or reference for graduate students in the field of integrated circuit fabrication. Major areas of device physics and processing are described and illustrated with simulations.

The material in this book is a result of several years of work on the implementation of the simulation system, the refinement of physical models in the simulation programs, and the application of the programs to many cases of device developments. The text began as publications in journals and conference proceedings, as well as lecture notes for an Hewlett-Packard internal CAD course.

This book consists of two parts. It begins with an overview of the status of CAD in VLSI, which points out why CAD is essential in VLSI development. Part A presents the organization of the two-dimensional simulation system. The process, device and parasitics simulation programs and application examples are presented. These chapters are intended to introduce the reader to the programs. The program structure and models used will be described only briefly. Since these programs are in the public domain (with the exception of the parasitic simulation programs), the reader is referred to the manuals for more details. In this second edition, the process program SUPREM III has been added to Chapter 2. The device simulation program PISCES has replaced the program SIFCOD in Chapter 3. A three-dimensional parasitics simulator FCAP3 has been added to Chapter 4. It is clear that these programs or other programs with similar capabilities will be indispensible for VLSI/ULSI device developments.

Part B of the book presents case studies, where the application of simulation tools to solve VLSI device design problems is described in detail. The physics of the problems are illustrated with the aid of numerical simulations. Solutions to these problems are presented. Issues in state-of-the-art device development such as drain-induced barrier lowering, trench isolation, hot electron effects, device scaling and interconnect parasitics are discussed. In this second edition, two new chapters are added. Chapter 6 presents the methodology and significance of benchmarking simulation programs, in this case the SUPREM III program. Chapter 13 describes a systematic approach to investigate the sensitivity of device characteristics to process variations, as well as the trade-offs between different device designs. Materials on the simulation of parasitic capacitance and resistance (originially Chapter 13 in the first edition) now include both two-dimensional and three dimensional simulations, and are presented in Chapter 15.

Part B is organized as follows. Chapters 5 and 6 describes the basic methodology in process and device simulations. Chapters 7 to 11 discuss the issues in VLSI device development. After understanding these issues, the readers will be ready to consider device designs, which are described in Chapters 12 to 14. Chapter 15 shows examples of parasitics extractions.

For the book to be used as a textbook, we recommend that it be used for a semester course. If the course deals with device modeling and computer-aided design tools, then Part A of the book should be emphasized. The student will learn about the fundamentals of process and device simulation programs and simulation system organization. If the course deals with device

Preface

physics and process development, then Part B of the book should be emphasized. The student will learn about current issues in VLSI device development. In either case, Part A and Part B of the book will compliment each other.

A note about the symbols used in this book. The symbols used in different simulation programs are often not consistent with each other. Therefore, the reader may find that symbols in this book not standardized throughout the chapters. We appologize for this confusion.

We are grateful that Dr. John Chi-Hung Hui has contributed Chapter 9 on the issue of hot electron degradation effects in submicron n-channel MOSFETs. The optimization of the LDD structure for reducing the hot electron degradation is described in detail.

We are also grateful that Dr. Sukgi Choi has contributed Chapter 14 on the issue of device scaling. The scaling of n-channel enhancement and depletion mode devices are presented. Factors causing the device characteristics to deviate from classical scaling rules, as well as complications involved in short channel device scaling such as punchthrough are discussed.

We are indebted to Dr. D. Wenocur and Mr. M. Varon for proof reading the manuscripts for the first edition. Mr. K. Okasaki has been assisting us with the computer systems which produced many of the figures in this book, as well as the camera-ready manuscript. He is also reponsible for debugging and modifying the simulation programs. Thanks are to Mr. T. Ekstedt who has kindly assisted the formatting of the first edition, and Dr. S.-L. Ng who has assisted in preparing many of the figures.

We are indebted to many of our colleagues at Hewlett-Packard Laboratories for providing us with many ideas and suggestions, and to our management for providing an opportunity for us to complete this task.

Finally, we like to thank our families for their spiritual support, patience and understanding during the preparation of the manuscript.

COMPUTER-AIDED DESIGN AND VLSI DEVICE DEVELOPMENT
SECOND EDITION

Overview

In order to bring out the importance of Computer-Aided Design (CAD) in VLSI (Very-Large-Scale Integration) device design, it is necessary to discuss the evolution of the Metal-Oxide-Semiconductor Field-Effect Transistors (MOSFETs) and the issues involved in its scaling. MOSFETs, first proposed 50 years ago, are based on the principle of modulating longitudinal electrical conductance by varying a transverse electrical field. Since its conception, MOSFET technology has improved steadily and has become the primary technology for large-scale circuit integration on a monolithic chip, primarily because of the simple device structure. VLSI development for greater functional complexity and circuit performance on a single chip is strongly motivated by the reduced cost per device and has been achieved in part by larger chip areas, but predominantly by smaller device dimensions and the clever design of devices and circuits.

A general guide to the smaller devices in MOSFETs and associated benefits, has been proposed by Dennard et al [1] (MOSFET scaling). This proposed scheme assumes that the x and y dimensions (in the circuit plane) are large compared to the z-dimension for the active device. The scaling method is also restricted to MOS devices and circuits. The active portion of MOS devices is typically restricted to within one or two microns of the crystal surface and interconnection dielectrics and metals are less than one micron in thickness. Thus we should expect that the guidelines as proposed by Dennard should be

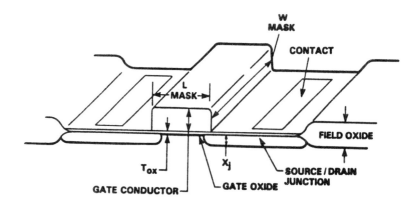

Fig. 1. Generic MOS Transistor.

reasonably valid for minimum circuit dimensions of two microns or greater. As the dimensions decrease below two microns, problems are introduced in both fabrications and device operations that are not significant in larger long-channel devices. The 2-D aspects of the impurity profiles and oxidation process become important in determining the effective channel length and width. More processing steps are required, such as channel implantation and local oxidation, which make more stringent control of the process necessary. Secondary effects, such as oxidation-enhanced diffusion significantly affect the impurity profile. As a result, better understanding and accurate control of these phenomena are crucial to achieving the desired performance from the scaled devices.

For the device operation, we will examine many of the scaling assumptions as applied to the long channel, wide conductor circuits. Also, in each case, there have been practical departures from the scaling assumptions. The original rules proposed that physical dimensions were scaled so that all electric field patterns were kept constant. Fig. 1 shows a generic MOS transistor and the various dimensions. Table 1 gives Dennard's constant field scaling rule, even though these have not been the general practice. When a process is scaled to smaller dimensions, the x, y and z dimensions are all scaled by the same amount. In addition, the applied voltages are scaled in constant field scaling to maintain constant field pattern.

W, L, T_{ox}, V, N_b		$\alpha\, K^{-1}$
I_{DS}	$\alpha\, (W/L)(V^2/T_{ox})$	$\alpha\, K^{-1}$
C_g	$\alpha\, WLC_{ox}$	$\alpha\, K^{-1}$
t_d	$\alpha\, C_g V/I$	$\alpha\, K^{-1}$
P	$\alpha\, VI$	$\alpha\, K^{-2}$
P/A	$\alpha\, VI/WL$	$\alpha\, 1$
Pt_d		$\alpha\, K^{-3}$

Table 1. Dennard's constant field scaling.

Many scaling schemes have been proposed since the "constant field" proposal. It is useful to consider the actual scaling methods that have been followed. The desirability of electrical compatibility with bipolar TTL circuits and the five volt power supply standard has resulted in "constant voltage scaling" as far as circuit power supply is concerned. Sometimes the internal node voltages are changed as a result of scaling and re-design. If the source/drain junction is less than one micron and V_{DD} is retained at five volts, the electric field stress in the channel is too great and the MOS transistor characteristics drift with time because of the hot carrier charge trapping. This constant voltage scaling also makes the 2-D field coupling significant, which is negligible in the long-channel device. It is the major cause of all the short-channel effects. To model these short-channel effects, 2-D numerical simulations become necessary because 1-D analytical models are not adequate. We must, in any event, consider lower system voltage at some future time. Following examples illustrate several features of the scaling methods. The most evident is that whereas most features of Dennard's constant field scaling are approximately retained from one generation of technology to the next, practical considerations have resulted in significant departures. Constant voltage scaling has in fact been the primary mode of scaling for most merchant suppliers. We can expect that a new power supply standard will be adopted and used until the dimensions are once more so small that device instability re-appears. Another departure from strict geometrical scaling has been in the vertical thickness of films. Conductor thickness has scaled very slowly in order to avoid electromigration effects in aluminum, or signal delay effects in polysilicon conductors. Table 2 shows the actual scaling done by most industrial suppliers. Either constant voltage or constant field scaling has resulted in improved circuits as measured by speed, and chip size and power for a given electronic function. As was mentioned earlier, circuit voltage, V_{DD}, will almost certainly be reduced for sub-micron devices. The tendency not to scale the interconnect or dielectric thickness (except gate oxide) will continue.

The future reduction of minimum features to less than one micron will undoubtedly bring further changes in actual scaling effects. The minimum practical conduction threshold for switches to turn off in dynamic circuits is approximately 0.6 volts. The desirability of dynamic operation in many electrical functions will keep CMOS circuit operation at greater than two volts. There will be exceptions such as the watch circuit that operates from a single battery cell. The peak current ($V_{GS} = V_{DS}/2$) per unit width scales as K^2 for no velocity saturation and constant voltage scaling. The effect of velocity saturation is to reduce this scaling factor to approximately K. If width is also reduced by the same scaling factor of K, then peak current per unit width scales as K for long channel and is almost constant for short channel. The transconductance follows the same behavior as current. Some switches such as the transistors in a static RAM cell are not required to switch particularly fast and have a small capacitive load. Hence the relative lack of increased performance from scaling, the smallest geometry device does not result in a performance problem inside the static RAM cell. On the other hand, devices that must drive long signal lines, clock lines, or word lines may require width-scaling that is not at all the same as length-scaling.

The scaling for commercial devices then will be to a new voltage standard of less than five volts as dimensions become sub-micron. The tendency in node capacitance is such that average wire length is a fraction of chip size. As

PROCESS (COMPANY)	Leff (um)	Vt (volt)	VDD (volt)	Tox (nm)	Xj (um)
NMOS(Intel)	4.6		5.0	120	2.00
NMOS(HP)	3.0	0.8	5.0	100	
HMOSI(Intel)	2.9	0.7	5.0	70	0.80
NMOS(Xerox)	2.5		5.0	70	0.46
NMOS(HP)	2.0	0.8	5.0	50	0.20
HMOSII(Intel)	1.6	0.7	5.0	40	0.80
NMOS(HP)	1.4	0.6	3.0	40	0.30
HMOSIII(Intel)	1.1	0.7	5.0	25	0.30
NMOS(NTT)	1.0	0.5	5.0	30	0.25
NMOS(IBM)	0.8	0.6	2.5	25	0.35
NMOS(Toshiba)	0.5	0.5	3.0	15	0.23
NMOS(AT&T)	0.3		1.5	25	0.26

Table 2. Actual scaling done by the most industrial suppliers.

more electronic functions are included on a chip, the chip size will continue to increase, and wire length also increases. In future scaling, the capacitance per unit length of wire for minimum pitch will stay almost constant since a large part of capacitance is fringing field or else inter-line. A new circuit design problem will be wire placement to minimize capacitance effects. To evaluate the actual capacitance values, the circuit extraction program coupled with the 2-D or 3-D parasitics simulator is indispensable.

An additional feature of scaling will be that the logic device width will be reduced by a smaller factor than the scaling factor. This feature is a result of the fact that the driving current per unit width only scales at best as the loading capacitance, $C_o \approx K$. If the width is scaled down by a factor of K, then the driving current per device at constant voltage will decrease somewhat as a result of various parasitic effects. The examples of minimum operating voltage and deleterious effects on performance are given to help illustrate the practical approach to sub-micron scaling.

The ability to calculate the effect of a process change on circuit or device electrical parameters has been an indispensable part of the rapid advances that have been made in semiconductor circuits ever since the beginning of the solid state micro-electronic industry. Experiments tend to establish the validity of theoretical concepts, establish empirical laws, as well as help discovering new physical effects. A symbiotic relation has developed between experiment and associated theory (modeling) in which the modeling helps to guide the direction of experiments, and experiments establish the validity of models as well as produce devices and circuits with optimized performance.

The conventional process and device designs for integrated-circuit technologies have been based on a trial-and-error approach using fabrications and measurements plus simple 1-D analytical modeling to achieve the desired terminal electrical characteristics and circuit performances. The left half of Fig. 2 outlines a process, device, and circuit design using the fabrications and simple models. This approach is not, however, adequate for the small geometry devices. As mentioned in the review of scaling, the constant voltage scaling and the lack of vertical scaling make the 2-D field coupling more dominant in the device performance. Especially, the threshold voltage becomes a sensitive function of the channel length and the drain bias. The fringing and inter-line capacitances become significant in the wiring capacitance. The velocity saturation also prohibits the simple 1-D model from accurately predicting the saturation currents. These factors force engineers to resort more to experiments. Thus, it drastically increases cost and time to develop a scaled geometry process. Even with the experiments, complicated processes and structures make it difficult to get physical insight and quantitative analysis of the factors governing device operation.

A complementary analysis and design path through process, device, and circuit simulations has been proposed and is now widely accepted. In process simulation, process-specification information is used to simulate the device structure and impurity distribution using the process models. Device simulation

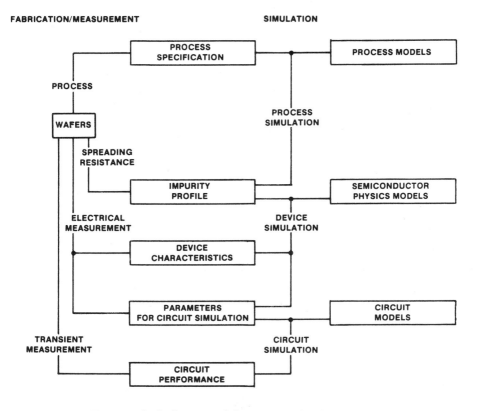

Fig. 2. Block diagram of the process development.

yields terminal characteristics based on the device structure and the impurity profile from the process simulation and physical models. The SPICE parameters are extracted from terminal characteristics and layouts. Based on these parameters and circuit connectivity, circuit simulations yield switching characteristics and provide means to evaluate the circuit performances. Compared to laboratory experiment, the design path via simulation is less costly and faster; more important, it produces detailed information concerning device operation in a well-controlled environment.

A complete 2-D numerical simulation system has been implemented in Hewlett Packard Laboratories since 1982. In part A of this book, this numerical simulation system and its individual tools will be explained in detail so that the reader can get acquainted with these tools and learn how to use them. Most of these tools are in the public domain. Thus we also give the information of these programs in appendix so that it will help readers implement these tools. In part B, the applications of the system in modeling small geometry processes will be presented. These simulation tools are different from the simple analytical models. First, we try to develop a methodology to use these tools effectively in process development. Next, real examples which are typical in important topics of scaled process development will be given in detail to help readers attack their real problems.

As can be seen through the examples, Computer-Aided-Design has been found to be absolutely essential in VLSI process and device development. As structures on the integrated circuit, such as transistors, field isolations, and interconnects are scaled down in size, many issues which have not been important have become significant. In case of the transistor, examples of these issues are the drain-induced barrier lowering, and hot electron effects. For interconnects, the issues are fringing-field capacitance and interline coupling. The physics of these issues cannot be understood in simple terms. Numerical simulations are the best tool to solve these problems. As a generality, we cannot expect totally accurate values of all physical parameters from simulations. Diffusion and oxidation processes are such that second order phenomena can be dominant. The result is that equivalent circuit parameters for SPICE or other simulations cannot be calculated with great accuracy just from process variables. Generally the changes in parameter values are more accurately predicted. The shortcoming of simulation is not fatal in regards to utility. The number of experiments that are required to achieve an optimum result is still greatly reduced by using simulation. It is also possible to reach an optimum design by using the ability to calculate the change of parameters with process variation.

It is our experience that calculation of resistance, capacitance, and inductance in interconnect lines is dependable and accurate. The long channel conduction threshold is also quite accurate. Short or narrow channel conduction threshold shows the correct trend but is not as accurate as needed for circuit usage. Also the conduction saturation current must be empirically adjusted. For these reasons of necessity to empirically adjust various physical parameters, we should always include appropriate test structures in test masks. With appropriate caution, it is possible to supply typical device parameters as well as parametric spread to circuit designers at an early stage in a process development.

There are many device properties that benefit from simulation capabilities, but do not appear in an equivalent circuit. These characteristics are, for example, drain to source punchthrough voltage; drain (or source) avalanche voltage; substrate or gate hot carrier current. These characteristics are often indirectly specified in terms of supply voltage, maximum current density of interconnect, or other limiting design rule. These characteristics are also often very difficult to simulate accurately. The ability to use two or three dimensional simulation is still of great use in designing devices with proper limits on these indirectly applied rules.

A further use of simulation is in understanding the internal device physics. That is, simulations allow one to observe quantities such as potential and electric field distributions, which are not measurable directly. As shown in Part B of this book, the knowledge of the internal properties of the devices is essential to the understanding of the observed device behavior. Good examples are the drain-induced barrier lowering (Ch. 8) and the hot carrier effect (Ch. 9). In the case of interconnection capacitance, the potential profiles generated by the simulator allow one to understand better the capacitive coupling between the conductive layers (Ch. 15). As device and interconnection structures become more complicated, it is almost impossible to predict or understand their characteristics without the help of numerical simulators.

Reference

[1] R. H. Dennard, F. H. Gaensslen, H. N. Yu, V. L. Rideout, E. Bassous, and A. Le Blanc, "Design of Ion-Implanted MOSFET's with very Small Physical Dimensions," *IEEE J. Solid-State Circuits*, **SC-9**, Oct 1974, pp. 256-268.

PART A
Numerical Simulation Systems

Chapter 1
Numerical Simulation Systems

1.1 History of Numerical Simulation Systems

Historically, numerical simulations of the MOS device have been used first to understand the device operation in the subthreshold and saturation regions. In 1969, Barron [1.1] from Stanford University simulated a MOSFET transistor using a finite-difference method to study the subthreshold conduction and saturation mechanism. Vandorpe [1.2] also simulated and modeled the saturation region with the finite-difference program in 1972. After the self-aligned silicon gate technology was invented, MOSFET device dimensions were reduced. This reduction prompted more numerical simulations to study the short-channel and narrow-width effects. Mock and Kennedy [1.3] from IBM developed a finite-difference program. Hatchel [1.4] also from IBM developed the first finite-element device simulation program. Barnes [1.5] from University of Michigan also developed a finite-element device simulation program for GaAs MESFETs. Most of the programs mentioned above were developed as research tools rather than as design tools for the general users. More emphasis had been put on the development of a stable and fast algorithm and the implementations of the physical mechanisms rather than on the user interface.

As MOS devices have been shrunk further, several groups started to develop the 2-D device simulator as a general design tool for the small geometry devices. One group was Cottrell/Buturla [1.6] from IBM who developed the finite-element program FIELDAID. Mock [1.7] also developed the finite-difference program CADDET using the stream function with Toyabe from Hitachi. Greenfield [1.8] and Dutton from Stanford University developed the finite-difference 2-D Poisson solver, GEMINI, which is only valid for the subthreshold and linear regions. Selberherr and Potzl [1.9] from University of Vienna developed 2-D 1-carrier program, MINIMOS using the finite-difference method. With the exception of FIELDAID, these programs are limited in input geometries and simulate the steady-state case. They are sufficient for the conventional structure MOS devices. However, as the device structures become more complicated and the speed improves, it becomes necessary to have the capabilities to simulate the arbitrary-shape geometry and the transient phenomena. Recently, Mock [1.10] has developed the SIFCOD program to satisfy these needs. SIFCOD is a 2-D 2-carrier device simulator with the arbitrary-shape geometry and transient device simulations including the small number of lumped circuit elements. In 1984, Pinto [1.11] from Stanford University has also developed the PISCES II program, which is a 2-D 2-carrier device simulator with arbitrary-shape input. It can simulate the steady-state and transient cases.

Process simulation started much later compared with device simulation. In the 60's and 70's, most of process modelings had been done by the analytical equations such as the diffusion equation and Deal-Grove oxidation equation[1.12]. As the process became sophisticated and the device was scaled down, the secondary effects such as the doping dependence of diffusivity, oxidation enhanced diffusion, severely affected the accuracy of the simple analytical model. In 1977, Antoniadis and Dutton [1.13] from Stanford introduced the SUPREM program, which is a 1-D numerical process simulator. It became very popular and an indispensable tool in process development.

As the device shrinks further, the 2-D effects are increasingly important in the process modeling as in the device modeling. Lee [1.14] from Stanford University developed 2-D process simulator, BIRD, which employed Green's function to solve the diffusion equation. The Green's function approach requires constant diffusivity and is only valid in the low concentration case. To overcome this limitation, Chin and Kump [1.15] also from Stanford University developed a numerical 2-D process simulator, SUPRA using the Green's function method for low impurity concentration as in BIRD and finite-difference method for high impurity concentration.

1.2 Implementation of a Numerical Simulation System

As mentioned in the previous chapter, many 2-D simulation programs have been developed and several have been reported recently. Most of these programs are more research tools rather than design tools. More emphasis has been put on the development of fast algorithms and the implementation of the physical mechanism than on the user interface. Futhermore, each program was developed independently and without an interface to other programs. Thus, transferring massive amounts of 2-D numerical data from one program to another is very difficult. Analyzing and interpreting the data is also difficult. To overcome these problems and provide a convenient design path using simulation, a complete 2-D simulation system has been developed at Hewlett Packard with an emphasis on the user interface. The following schemes have been adopted to make this system a more practical, user-oriented and well supported design tool.

1) *Mode of program execution* : To reduce the engineering time, users want to control the input preparation, program execution and data file handling. Ideally, users can run the programs in interactive mode for the quick check of the input and the graphical post processing and run in batch mode for the time consuming simulation job. Thus, all programs have been implemented on the mini-computer in our laboratory with an emphasis on the user control and fast turn-around.
2) *Hierarchical simulation* : Full approach for the 2-D simulations is time consuming. This is especially true in process and device simulations. Thus, simplified programs are implemented together with the full approach program so that users can select according to their application to save the simulation time.

3) *Interface between programs* : These simulation programs generate a large amount of 2-D data. A clean and user-transparent procedure is important to transfer data between the programs. Standard format of 2-D data file has been set up and utilized.
4) *Graphical post processing* : To manipulate and analyze the massive 2-D data graphical post processing is absolutely essential. A plot package has been implemented which can do the 3-D bird's-eye-view plot, 2-D contour plot, and 1-D cross-section plot all together.
5) *Complete system support* : For better user acceptance, system support in the form of the training and manual is a critical issue. Users are usually afraid to read the manual and use programs without training because they may spend too much time without any results. Ideally, training can be means for users to become acquainted with and begin to use a program. Thereafter, the manual can be used for reference. The manual should be easy to understand, complete and up-to-date. An on-line manual is very helpful.
6) *Easy to enhance* : The physical models in the process and device simulation are still evolving. Sometimes, the simulation programs are used as development vehicles for new models. Thus, it is necessary to make the programs modular so that they are easy to modify and repair.
7) *Bench-mark of the simulation tools* : Without bench-marking, the tools are useless. It is necessary either to bench-mark the tools in a wide range of the practical interest or specify the range where the tool is valid.

A block diagram of the system is shown in Fig. 1.1. For process simulation, there are the SUPREM, SUPRA and SOAP programs. SUPREM is a 1-D process program capable of simulating most typical IC fabrication steps such as deposition, etch, ion implantation, diffusion and oxidation. It generates the 1-D profile of all the dopants present in the silicon and silicon dioxide. SUPREM is ideal for simulation of the doping profile of the bipolar or MOS capacitor where the 1-D structure is dominant. The doping distribution of a MOS transistor is basically two dimensional. However, SUPREM can be used to generate the doping profile of MOS devices where the shape of 2-D source/drain lateral diffusion is not so critical. In this case, two SUPREM 1-D simulations are needed; one is to specify the doping distribution in the channel

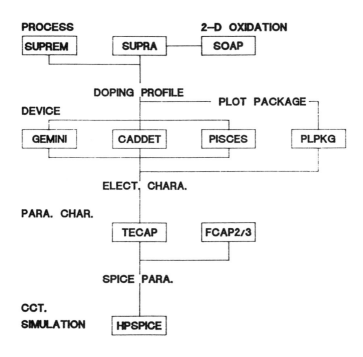

Fig. 1.1. Block diagram of numerical simulation system.

and the other is to specify the distribution in the source/drain region. The lateral diffusion is calculated by multiplying the vertical source/drain profile with Gaussian or complementary error functions. SUPRA is needed for more accurate 2-D doping profile in MOS transistors. The 2-D process program, SUPRA, simulates processes based on the device geometry and process schedule and generates the impurity distributions in two dimensions. SUPRA can handle deposition, etch, ion implantation, diffusion and oxidation process cycles. For impurity diffusions, SUPRA analytically solves the diffusion equation using a constant diffusivity for low impurity concentrations. For higher concentrations, it solves the diffusion equation using a concentration-dependent diffusivity in a numerical finite-difference method. The oxidation model is based on the measurements of the common semi-recessed oxidation. For a more accurate or exotic oxidation simulation, SOAP should be used. SOAP is a program that simulates the diffusion of the oxygen in the oxide and the propagation of the extra oxide volume generated during the oxidation in a rigorous manner. SOAP can be used together with SUPRA to simulate 2-D process including oxidation or in a stand-alone mode just for 2-D oxidation structure. Two-dimensional oxidation is a difficult problem to simulate because of nonplanar geometries and moving boundaries. SOAP uses the boundary-value method. In this method, the nodes are allocated only along the boundary. Thus, it is very suitable for nonplanar geometry and moving-boundary problems.

Electrical device characteristics are predicted by 2-D device simulators based on the impurity distributions predicted by process simulators. In the semiconductor device, the Poisson, electron and hole continuity equations should be solved with the appropriate boundary conditions. In MOS devices, however, most current is carried by the majority carrier of the source/drain. Thus, MOS devices can be simulated by only solving the majority carrier continuity equation. Such an algorithm is used by the full 2-D simulator CADDET. Thus, it can simulate the whole range of the MOS device operation. Due to the simplification, it cannot, however, simulate the phenomena related to the two carrier problem. The simultaneous solution of 2-D Poisson and current continuity equations makes CADDET slow. Simplified analysis should be used whenever possible. CADDET also has limitations in the input structure. For the subthreshold region, where the current is small, it is enough to solve Poisson's equation with proper guessing of the electron and hole quasi-Fermi level. GEMINI adopts this scheme and speeds up the simulations in the subthreshold region. It has good input and output capability and is fast, but care should be taken to use it within its limits. CADDET and GEMINI have some limitations due to the simplifications. These limitations are acceptable for many MOS device applications. However, they are fatal in some cases, especially in the hot electron related problem, latch-up, or novel devices. To satisfy these applications, a new general-shape 2-D 2-carrier transient device simulator, PISCES has been acquired and enhanced. It can simulate the steady-state device characteristics and the switching characteristics together with the lumped circuit elements. It accepts any shape geometry and solves for both carriers so that it can simulate any semiconductor device with almost no limitations.

Based on the device characteristics calculated by these simulators, the electrical parameters can be extracted by TECAP2 [1.16] for use in circuit simulations. TECAP2 measures the device characteristics of MOS transistors and extracts their parameters. Here, the data is taken from the simulations and only the parameter extraction part of TECAP2 is used. In VLSI circuits, accurate determination of interconnect and other capacitance values becomes crucial in the circuit simulation. The value of various capacitance components of MOS circuits can be simulated and determined by FCAP2/3 [1.17], a two- and three-dimensional, arbitrary-geometry, linear Poisson solver. The circuit performance can be simulated by HP-SPICE [1.18] based on electrical device parameters and capacitances that were obtained from the process schedule and device and circuit layouts using this simulation system. Each program will be explained in more detail in the subsequent chapters. The process simulators will be discussed in chapter 2 and the device simulators will be in chapter 3. Chapter 4 will deal with parasitics simulation. TECAP2 and SPICE will not be treated because they are not closely related to the topics in this book. The purpose of part I is to introduce these programs to the users as simply as possible so that they can easily understand how these programs work and give them useful examples so that they can start to simulate the practical problems by just modifying the examples without spending too much time to read manuals. The main reason people do not use programs is the bulkiness of the manual. Once a user is familiar with running the program, he can build up his skill and knowledge with practice and by using the manual. The necessary things for the users to know about each program (its organization, numerical algorithm, and physical model) are briefly explained. After that, practical examples are given to illustrate how to use the program. First, input example files are explained in detail with a brief illustration of the input structure. Common pitfalls are also illustrated to prevent the users from spending too much time without any results. Then the simulated results such as impurity or potential profile are presented and discussed briefly.

References

[1.1] M. B. Barron, "Computer Aided Analysis of IGFET Transistor," TR No.5501-1, Stanford Electronics Laboratories, Stanford University,

Stanford, CA, Nov 1969.

[1.2] D. Vandorpe and J. Borel, "An Accurate Two-Dimensional Numerical Analysis of the MOS Transistor," *Solid State-Electronics*, **15**, 1972, pp 547-557.

[1.3] M. S. Mock, "A Two-Dimensional Mathematical Method of IGFET Transistors," *Solid-State Electronics*, **16**, 1973, pp. 601-609.

[1.4] G. D. Hatchel et al, "A Graphical Study of the Current Distribution in Short-Channel IGFETs," *ISSCC Conf. Digest*, Philadelphia, pp. 110-111, 1974.

[1.5] J. J. Barnes and R. J. Lomax, " Two-Dimensional Finite Element Simulation of Semiconductor Devices," *Electron. Lett.*, **10**, 8 Aug 1974, pp. 341-343.

[1.6] P. E. Cottrell and E. M. Buturla, " Steady-state Analysis of Field Effect Transistors via the Finite Element Method," *IEDM Tech. Digest*, Dec 1975, pp. 51-54.

[1.7] T. Toyabe, K. Yamaguchi, S. Asai, and M. S. Mock, "A Numerical Model of Avalanche Breakdown in MOSFETs," *IEEE Trans. on Electron Devices*, **ED-25**, July 1978, pp. 825-831.

[1.8] J. A. Greenfield, S. E. Hansen, and R. W. Dutton, "Two-Dimensional Analysis for Device Modeling," TR No. G201-7, Stanford Electronics Laboratories, Stanford University, Stanford, Calif., 1980.

[1.9] S. Selberherr, W. Fichtner and H. W. Potzl, "Minimos - A Program Package to Facilitate MOS Device Design and Analysis," *Proc. of NASECODE I Conf.* June, 1979, pp. 275-279,

[1.10] M. S. Mock, "A Time-Dependent Numerical Model of the Insulated-Gate Field-Effect Transistor," *Solid-State Electronics* **24**, pp. 959-966, 1981.

[1.11] M. R. Pinto et al, " Computer-Aids for Analysis and Scaling of Extrinsic Devices," *IEDM Tech. Digest*, Dec 1984, pp. 288-291.

[1.12] B. E. Deal and A. S. Grove, *J. Appl. Physics.*, 36, 1965, p. 377

[1.13] D. A. Antoniadis, S. E. Hansen, and R. W. Dutton, "SUPREM II - A Program for IC Process Modeling and Simulation," TR No. 5019-2, Stanford Electronics Laboratories, Stanford University, Stanford, Calif., 1978.

[1.14] H. G. Lee, "Two-Dimensional Impurity Diffusion Studies: Process Models and Test Structures for Low-Concentration Boron Diffusion," TR No. G201-8, Stanford Electronics Laboratories, Stanford University, Stanford, Calif., Aug 1980.

[1.15] D. Chin, M. Kump, and R. W. Dutton, SUPRA: Stanford University Process Analysis Program," Stanford Electronics Laboratories, Stanford University, Stanford, Calif., Oct 1979.

[1.16] E. Khalily, "Transistor Electrical Characterization and Analysis Program," *Hewlett-Packard Journal*, **Vol. 32, No. 6**, June 1981.

[1.17] Soo-Young Oh, " MOS Device and Process Design Using Computer Simulations," *Hewlett-Packard Journal*, **Vol. 33, No. 10**, Oct 1982.

[1.18] L. K. Scheffer, R. I. Dowell, and R. M. Apte, "Design and Simulation of VLSI Circuits," *Hewlett-Packard Journal*, **Vol. 32, No. 6**, June 1981.

Chapter 2
Process Simulation

2.1 Introduction

Silicon integrated circuit (IC) technology has evolved to fabricate multi-million transistors on a single chip. Trial-and-error methodology to optimize such a complex process is no longer desirable because of the enormous cost and turn-around time. From this point of view, computer simulation is a cost-effective alternative, not only supplying a right answer for increasingly tight processing windows, but also serving as a tool to develop future technologies. When coupled with a device analysis program, a process simulator has proven to be a powerful design tool because the process sensitivity to device parameters can be easily extracted by simple changes made to processing conditions in computer inputs [2.1].

SUPREM (Stanford University PRocess Engineering Models) [2.1] was introduced in 1977 and was the first program capable of simulating most IC fabrication steps. As of 1983, more than 300 copies were distributed world wide and enhanced versions SUPREM II [2.2] and SUPREM III [2.3] are now available. The program accepts a process-runsheet-like input and gives an output containing the impurity distributions in the vertical direction. SUPREM, therefore, can be applied to any regions where impurity distribution changes only in the vertical direction as indicated in a CMOS cross-section (Fig. 2.1). Fabrication of the structure involves several processing steps of ion implantation, oxidation/drive-in and etching/deposition.

The SUPREM program consists of various models based on experimental data as well as physical assumptions which will be discussed in detail in following sections. Some of the models, however, have severe limits in their valid ranges and some are not even fully understood; phosphorus diffusion is an example of this. Parameters associated with these models may also be subjective to individual processing conditions. Users may find occasional discrepancies between measured data and simulation results if they simply use default values in the program. It should be remembered that users need to check whether a model and associated default values in SUPREM are valid for process steps to be simulated. If they are not valid, parameters should be adjusted in order to obtain more accurate results.

With the trend toward shallow junctions and lower heat cycles in VLSI technologies, two-dimensional impurity profiles and structures are more crucial to device characteristics. Threshold voltage and parasitic capacitance, for example, are strong functions of lateral diffusion of arsenic in the source/drain and boron in the channel-stop region. Simple extension of SUPREM to two dimensions may not be desirable because the degree of equivalent numerical accuracy requires tremendous computing resources. Furthermore, device structures change continuously during such processes as reactive-ion etching and local oxidation of silicon (LOCOS), which impose more difficulties to establishing simulation algorithms. SUPRA (Stanford University PRocess Analysis) [2.4] introduced in 1981 was one of the pioneer works in the two-dimensional process modeling and it can be applied to the areas where impurity distributions and device structures change not only in the vertical direction but in the lateral direction as indicated in Fig. 2.1. It had already been demonstrated that the two-dimensional process simulator could be a powerful design tool when coupled with a device analysis program [2.5]. Nonuniform processing sometimes needs a new concept to explain the physical mechanism. The local oxidation is an example that requires significantly different kinetic equations from the one-dimensional counterpart. SOAP (Stanford Oxidation Analysis Program) [2.6] in 1983 is a program simulating nonuniform oxidation

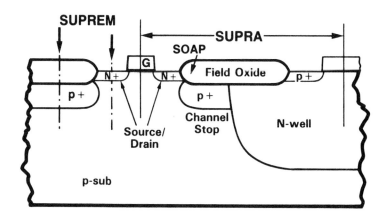

Fig. 2.1. Cross-section of a CMOS device.

processes.

There may be many other tools [2.7-9] developed by companies or universities but used only internally or not publicized. In this chapter, we will discuss SUPREM, SUPRA, and SOAP because they are available for public use. The following sections describe physical models and application examples of the programs.

2.2 SUPREM: 1-D Process Simulator

There are two generations of the SUPREM program that are widely used: SUPREM II, first released in 1978, and SUPREM III, released in 1983. Each program has gone through several modifications and upgrades since their first release. SUPREM III is a substantial enhancement of SUPREM II and allows more layers, material types and dopant types than SUPREM II. Table 2.1 summarizes some of the differences between the two programs. In general, SUPREM III uses more sophisticated process models than does SUPREM II.

Each program supports the following process steps: ion implantation, oxidation, chemical predeposition, diffusion, epitaxial growth, etch and deposition. The input syntax for the two versions of SUPREM are different. However, in both programs, the input file resembles an actual process runsheet consisting of free format statements involving key words and numbers.

SUPREM II	SUPREM III
2 layers	10 layers
default materials for implant and diffusion silicon silicon dioxide	default materials for implant and diffusion silicon silicon dioxide poly silicon silicon nitride aluminum
no additional material types	additional materials may be defined
3 dopant types	4 dopant types
default dopants boron phosphorus arsenic antimony	default dopants boron/BF_2 phosphorus arsenic antimony
oxidation of silicon	oxidation of silicon poly silicon silicon nitride

Table 2.1. Comparison of SUPREM II and SUPREM III

Process Simulation

Typically a single process step can be simulated with a single statement consisting of less than 60 alphanumeric characters.

When the program is executed it first goes through the input file, line by line, checking the validity of the syntax. Then each process step is executed sequentially as it appears in the input file. The output of the program, available at the end of each process step, consists of the one-dimensional profiles of all the dopants present in the various material layers.

Inside the program, the simulation structure is represented by a one dimensional line of grid points defined by the program user. Each grid point is labeled with a material type and up to three (SUPREM II) or four (SUPREM III) dopant concentrations. This array of grid points is modified to simulate the various process steps. Grid points are added to the top of the structure to describe epitaxial layers or new deposited layers. Likewise grid points are removed as material is etched away. Associated with each grid point is a cell whose boundaries lie midway between the grid points. Dopant can be added to the various cells by implantation. Dopant can be redistributed from one cell to another by diffusion. Dopant diffusion across interfaces is controlled by a segregation mechanism. When silicon is oxidized, the grid points near the top of the silicon layer gradually change material type from silicon to oxide. In this way the growing oxide consumes the top of the silicon. Oxidation and diffusion operations are divided into many time steps to accurately describe the time evolution of the simulation structure and dopant distributions.

Process Models

a) Ion Implantation

The implantation of ions into a material produces a peaked distribution as shown in Fig. 2.2. The gaussian function, Fig. 2.2a, is often used to describe the shape of the dopant profile.

$$C(x) = C_p \exp\left[-\frac{(x-R_p)^2}{2\sigma_1^2}\right] \quad (2.1)$$

Here C_p is the peak concentration. R_p is the position of the maximum of the

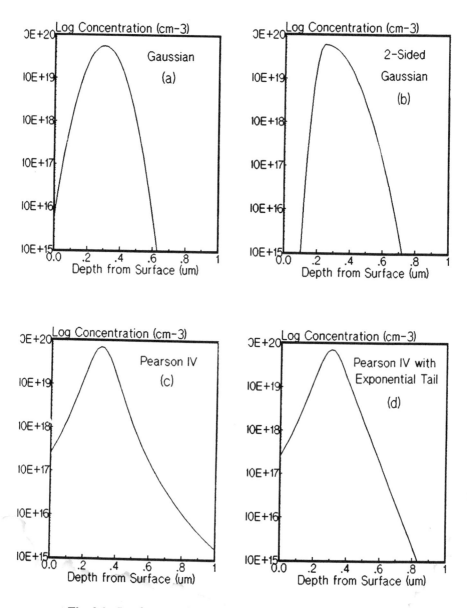

Fig. 2.2. Implant Profiles: (a) Gaussian (b) Two-Sided Gaussian (c) Pearson IV (moments chosen to model a distinct channeling tail) (d) Pearson IV with Exponential Tail.

distribution and is determined by the energy of the implantation. R_p is usually called the "projected range". σ is the standard deviation and determines the width of the distribution. Typically σ increases as R_p increases. The LSS theory [2.10] successfully predicts the standard deviation and projected range for most implantations into silicon.

The gaussian distribution is symmetric about R_p. However actual implant profiles often display some asymmetry or skewness. The joint half-Gaussian expression (Fig. 2.2b) is commonly used to describe this effect. Here two gaussian functions with different standard deviations are pieced together at the peak of the distribution.

$$C(x) = C_p \exp\left[-\frac{(x-R_p)^2}{2\sigma_1^2}\right] \quad 0 < x < R_p \qquad (2.2a)$$

$$C(x) = C_p \exp\left[-\frac{(x-R_p)^2}{2\sigma_2^2}\right] \quad R_p < x < \infty \qquad (2.2b)$$

Real dopant distributions may be more complicated than can be described with the joint half-gaussian. Some implant profiles display a "tail" due to channeling of the incident ions along the crystal planes. Boron implants display this property quite clearly [2.11]. This type of profile is illustrated in Fig. 2.2c,d. Such profiles could be approximated by a sum of gaussian functions. However, this is not commonly done.

Distribution functions can be characterized by their moments, μ_i.

$$\mu_1 = R_p = \int_{-\infty}^{\infty} x f(x) dx \qquad (2.3a)$$

$$\mu_i = \int_{-\infty}^{\infty} (x-R_p)^i f(x) dx \qquad (2.3b)$$

$$\int_{-\infty}^{\infty} f(x) dx = 1 \qquad (2.3c)$$

Here $f(x)$ is a normalized distribution function that describes the shape of the dopant profile. The first moment is just the projected range and is usually very near the peak of the distribution. The second moment is the square of the standard deviation, $\sigma = \sqrt{\mu_2}$, and describes the width of the distribution. The third moment accounts for the asymmetry (skewness) of the profile about R_p.

The forth moment contains most of the information about the tail of the distribution. Higher order moments describe the details of the function at distances farther from R_p.

The gaussian distribution is fully specified by the first two moments. The joint half-Gaussian is roughly equivalent to including the third moment. There exists a class of functions that are fully specified by the first four moments. These are called the Pearson distributions [2.12,13]. The Pearson type IV function is the only member of this class that is of interest for describing implant profiles. The Pearson IV distribution has the desirable properties of having a single maximum, no negative values and a smooth approach to zero for large and small values of x. The mathematical expression for the Pearson IV distribution is reasonably complex and is not presented here. Reference [2.14] provides a thorough discussion of this function. Pearson IV moments have been calculated from measured implant profiles for most common dopant species (boron, phosphorus, arsenic and antimony) [2.12,14,15]. Typically the results are summarized in a look-up table that lists the four moments as a function of implant energy.

One other approach has been used to describe the tail on dopant profiles particularly for boron implants. This involves splicing an exponential tail onto another distribution function such as gaussian or Pearson IV distribution [2.15]. The Pearson IV with an exponential tail is shown in Fig. 2.2d.

The LSS theory [2.10] of implantation yields gaussian profiles and predicts reasonable values for the range and standard deviation. Theoretical attempts to describe the deviations from simple gaussian profiles have been only partially successful [2.15]. The various functions used to describe implant profiles have been chosen simply because they offer an appropriate shape that can be fit to the measured data. Look-up tables of relative concentration versus depth for each dopant species and implant energy would offer more accurate descriptions of the measured profiles. However this is rarely done.

SUPREM II allows the program user to specify a Gaussian profile for any of the dopants. If the user does not specify a Gaussian profile, then the default is to use the two-sided Gaussian for arsenic, phosphorus and antimony. The Pearson IV with exponential tail is used for boron. The characteristic length of the exponential tail is fixed at 45 nm independent of implant energy.

The moments used in the two-sided Gaussian and Pearson IV profiles are not user definable.

SUPREM III offers the Gaussian, joint half-Gaussian and Pearson IV for each of the dopant types: boron, BF_2, phosphorus, arsenic and antimony. All of the moments used in these profiles may be adjusted by the program user. The default is to use the Pearson IV for all elements.

Both versions of SUPREM allow implants into multi-layer structures where each layer has different implant moments. SUPREM II, which is limited to two layers, has default implant parameters for silicon and silicon dioxide. SUPREM III, which can have up to ten layers, has implant parameters for silicon, silicon dioxide, poly-silicon, silicon nitride and aluminum.

The simulation structure is composed of a one-dimensional line of grid points and their associated cells. The implant statement causes a quantity of dopant to be added to each cell. This quantity is determined by the implant distribution evaluated at that grid point and the width of the cell. The amount of dopant in each cell is then scaled so that the total implant dose in all the cells of the various layers is equal to the specified dose. This assumes that the relative shape of the implant profile is independent of the implant dose which is usually a good assumption.

b) *Thermal Oxidation*

Silicon dioxidation (oxidation) is a thermal process in which oxidizing species diffuse through an oxide layer and react with silicon atoms. A 125 percent volume expansion accompanies the oxidation process. Three oxidant fluxes involved in the oxidation process are

$$F_1 = h(C^* - C_0) \tag{2.4a}$$

$$F_2 = D_{\mathit{eff}} \frac{C_0 - C_i}{X_0} \tag{2.4b}$$

$$F_3 = kC_i \tag{2.4c}$$

where
F_1 = transport flux from gas ambient to the oxide surface
F_2 = diffusion flux inside the oxide layer

F_3 = reaction flux at the silicon/oxide interface
C^* = equilibrium concentration in the oxide
C_0 = concentration at the oxide surface
C_i = concentration at the silicon/oxide interface
h = gas transport coefficient
D_{eff} = effective diffusion coefficient
X_0 = oxide thickness
k = surface reaction coefficient

The linear-parabolic oxide-growth model [2.16] in SUPREM assumes steady-state oxidant diffusion that the three fluxes are equal as

$$F_1 = F_2 = F_3 = F \qquad (2.5)$$

The oxide growth rate is directly proportional to the flux as

$$\frac{dX_0}{dt} = \frac{F}{N_1} = \frac{kC_i/N_1}{1 + k/h + kX_0/D_{eff}} \qquad (2.6)$$

where N_1 is the number of oxidant molecules incorporated in a unit volume of the oxide layer. When integrated, Eq. (2.6) leads to the well-known linear-parabolic growth relationship but only if an initial oxide X_i is specified prior to the oxidation step under consideration.

$$\frac{X_0^2 - X_i^2}{B} + \frac{X_0 - X_i}{B/A} = t \qquad (2.7)$$

where B is the parabolic rate constant and B/A is the linear rate constant. For relatively low dopant concentrations, B and B/A are independent of dopant levels and depend only on silicon crystal orientation, oxidizing ambient and temperature. The behavior of the two rate constants as a function of temperature are shown in Fig. 2.3. The default oxidation parameters in SUPREM are based on the data in this figure. The program user may have to adjust these parameters to fit the oxidation rates observed for a particular process and facility.

In SUPREM, oxidation operations are divided into many small time steps. The time steps are calculated by the program and are not user definable. SUPREM III provides limited user control over the minimum and maximum

Process Simulation

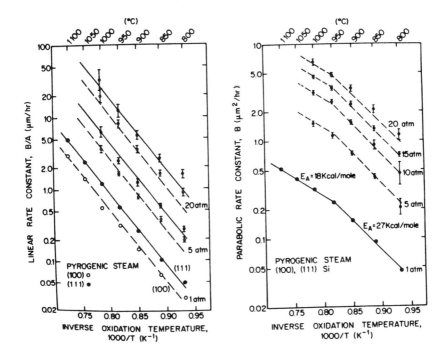

Fig. 2.3. Temperature dependence of oxide growth rates.

time step size. During the oxidation process the grid points and cells near the silicon surface are gradually transformed from silicon to oxide. During this transformation, the cells expand to give the correct volume expansion. The time step calculation is limited so that no more than one grid point is converted from silicon to oxide in a single time step.

The thin layer of silicon that is converted to oxide (in one time step) will normally contain some dopant. This dopant becomes part of the of the new oxide. After this is done, the effect of diffusion for that time step is computed. This allows the dopant to redistribute in the interface region.

SUPREM must calculate the amount of oxide grown, ΔX_0, during time step Δt. SUPREM II uses an incremental form of Eq. (2.7).

$$\Delta X_0 = \frac{1}{2}\left[-(2X_0+A) + \sqrt{(2X_0+A)^2 + 4B\Delta t}\right] \quad (2.8)$$

Here X_0 is the oxide thickness at the beginning of the time step. SUPREM III uses the differential form in Eq. (2.6).

$$\frac{\Delta X_0}{\Delta t} = \frac{B}{2X_0 + A} \quad (2.9)$$

The linear and parabolic rate constants are determined by the following equations in SUPREM III. The equations in SUPREM II are quite similar except that n, m, η and ε are all set equal to one. The temperature and pressure dependence of B/A and B is shown in Fig. 2.3 for low doping levels.

$$\frac{B}{A} = \left(\frac{B}{A}\right)_i P^n [1+\gamma(C_v+1)] \eta \alpha \quad (2.10a)$$

$$B = B_i P^m [1+\delta C_T^r] \varepsilon \quad (2.10b)$$

$\left(\dfrac{B}{A}\right)_i$ = linear rate constant for intrinsic silicon

B_i = parabolic rate constant for intrinsic silicon

P = oxidant pressure

C_T = total dopant concentration at Si/SiO2 interface

η and ε provide dependence on HCl concentration

α orientation dependence: $\alpha=1$ for <111> and $\alpha=0.6$ for <100>

C_v is the normalized vacancy density which is equal to one for intrinsic silicon and increases with doping concentration in extrinsic silicon. The terms involving C_v and C_T model the concentration enhanced oxidation effects [2.17]. The terms, η and ε, are determined from simple equations or look-up tables to model the enhanced or retarded oxidation due to chlorine in the oxidizing ambient.

In SUPREM, many parameters (B_i, $(B/A)_i$, γ, δ, etc.) are determined by an activation energy formulation. For example,

$$B_i = B_{i0} \exp\left[-\frac{E_B}{kT}\right] \quad (2.11a)$$

$$\gamma = \gamma_0 \exp\left[-\frac{E_\gamma}{kT}\right] \quad (2.11b)$$

and so on. These constants, B_{i0}, E_B, γ_0, E_γ, δ_0, E_δ, etc., are all user definable in SUPREM III. Only a limited subset of these parameters are user definable in

SUPREM II. Both dry and wet oxidations are modeled with the same set of formulas. However, the parameter values are quite different for the two types of oxidation.

The growth of thin oxides is difficult to model. It is known that the growth rate is enhanced as much as a factor of ten for oxides thinner than 20 nm grown in dry ambient. This phenomenon is quite important because the gate oxide in modern VLSI processes is grown within the thin oxide regime. To model thin oxide growth, SUPREM II artificially multiplies $(B/A)_i$ by 10 for $X_0 < 20$ nm. SUPREM III adds an empirical factor to the linear-parabolic model.

$$\frac{\Delta X_0}{\Delta t} = \frac{A}{2X_0 + B} + K \exp\left(-\frac{X_0}{L}\right) \qquad (2.12)$$

Here the decay length, L, is approximately 7 nm independent of temperature. K is a singly activated function of temperature [2.18].

c) *Impurity Redistribution*

In SUPREM, impurity diffusion during high temperature process steps is modeled with a basic continuity equation.

$$\frac{dC}{dt} = -\frac{dF}{dx} \qquad (2.13)$$

C is the concentration of a particular dopant species and F is the dopant current or flux. In SUPREM III, F is composed of a diffusion term and a drift term.

$$F = F_{diff} + F_{drift} \qquad (2.14)$$

The diffusion term is driven by the gradient of concentration. The drift term results from internal electric fields established by concentration gradients.

$$F_{diff} = -D \frac{dC}{dx} \qquad (2.15a)$$

$$F_{drift} = Z \mu E C^\dagger \qquad (2.15b)$$

C = chemical concentration

$C^\dagger$ = active concentration
D = diffusion constant
Z = charge state of active dopant
$\mu = \dfrac{qD}{kT}$ = mobility of active dopant
E = internal electric field

The electric field is determined by the gradient of the quasi-Fermi level.

$$E = -\frac{kT}{q}\frac{d}{dx}\left[\ln\left(\frac{n}{n_i}\right)\right] \tag{2.16}$$

n is the free electron concentration and n_i is the intrinsic electron concentration. SUPREM II does not include the drift term in Eq. (2.14). The effect of internal electric fields is approximated by a correction to the diffusion constant in the diffusion term.

Inside the program, the dopant concentration is known at each grid point. The dopant current is calculated at each cell boundary using the dopant concentration and carrier concentration at the two grid points on either side of the cell boundary. The gradient of the dopant current at a particular grid point is calculated using the dopant current at the two boundaries of the cell.

SUPREM utilizes models based on vacancy diffusion mechanisms under non-oxidizing conditions. The intrinsic diffusivity of an ionized impurity species is the sum of the diffusivities resulting from neutral vacancies and ionized vacancies with an opposite charge. There are four charged states for vacancies in silicon: double negative (=), singly negative (-), neutral (x) and positive (+). Thus, the effective diffusivity under non-oxidizing conditions is

$$D_N = D_i^x + D_i^-[V^-] + D_i^=[V^=] + D_i^+[V^+] \tag{2.17}$$

$[V^v]$ is the concentration of vacancies in each charge state, normalized to the intrinsic concentration of that state. The concentration of neutral vacancies at any given temperature is independent of the impurity concentration. The concentrations of charged vacancies depend on the Fermi level in the same way as the free electron concentration.

$$[V^-] = \frac{n}{n_i}, \quad [V^=] = \left(\frac{n}{n_i}\right)^2, \quad [V^+] = \frac{p}{n_i} \qquad (2.18)$$

Thus the charged vacancy density is a function of the doping density for extrinsic silicon (n or $p > n_i$). Under intrinsic conditions, the diffusion constant, D_i, is just the sum of the D_i^V's.

$$D_i = D_i^x + D_i^- + D_i^= + D_i^+ \qquad (2.19)$$

All the D_i^V's are singly activated functions of temperature. Not all the D_i^V's are significant for each dopant species. Acceptor atoms are negatively charged and diffuse primarily with neutral and positively charged vacancies. Conversely donor atoms diffuse mainly with neutral and negatively charged vacancies. In SUPREM III, the boron diffusivity is given by

$$D_N(B) = D_i^x + D_i^+ \left(\frac{p}{n_i}\right) \qquad (2.20)$$

and the diffusivities of arsenic and antimony are given by

$$D_N(Ar, Sb) = D_i^x + D_i^- \left(\frac{n}{n_i}\right) \qquad (2.21)$$

The formulation is slightly different in SUPREM II.

$$D_N = D_I \left[\frac{1 + \beta f_v}{1 + \beta}\right] \qquad (2.22)$$

Where $f_v = \frac{n}{n_i}$ for donors and $f_v = \frac{p}{n_i}$ for acceptors. Here again D_I is a singly activated function of temperature. β is equal to 3 for boron and equal to 100 for arsenic.

Both SUPREM II and III have special models for phosphorus diffusion. In the absence of large phosphorus concentration gradients, SUPREM III includes a contribution from doubly ionized vacancies.

$$D_N(P) = D_i^x + D_i^- \left(\frac{n}{n_i}\right) + D_i^= \left(\frac{n}{n_i}\right)^2 \qquad (2.23)$$

SUPREM II uses Eq. (2.22) with $\beta = 100$. For large phosphorus concentrations that are peaked near the silicon surface, the profile is divided into a surface region and a tail region [2.19]. The diffusion constant is increased in the tail region to model the "kink" observed in phosphorus profiles. The diffusion constant for all other dopants is also enhanced in the phosphorus tail region to account for the "base push" effect observed in bipolar devices.

It is well known that diffusion rates for most common dopants increase in oxidizing ambients. This is commonly known as oxidation enhanced diffusion or OED. At the present, this is thought to result from the injection of silicon self-interstitials from the oxidizing interface into the silicon bulk. Dopant atoms then have a larger probability of pairing with a silicon interstitial. The interaction of the dopant atom and interstitial provides an additional diffusion mechanism to the usual vacancy mechanism. SUPREM II models OED for phosphorus by artificially increasing the diffusion constant by a factor of 1.8 for dry oxidation and 3.3 for wet oxidation. OED is ignored for arsenic and antimony. Another term is added to the diffusion constant for boron in SUPREM II.

$$D(B) = D_N(B) + D_{ox} \qquad (2.24)$$

where D_{ox} is a singly activated function of temperature.

SUPREM III adds an additional term to the diffusion constant of each dopant type.

$$D = D_N + D_{ox} \qquad (2.25)$$

$$D_{ox} = D_I f_{ii} K \exp\left(-\frac{x}{L}\right) \left(\frac{dX}{dt}\right)^{0.5} f_{HCl} \qquad (2.26)$$

f_{ii} and K are activated functions of temperature. f_{ii} is specified differently for each dopant species while K is considered a property of the silicon. x is the distance from the oxide/silicon interface. L is the range of the interstitial diffusion which is usually quite large (approx. 25 um). dX/dt is the oxide growth rate. The function f_{HCl} provides some dependence on chlorine concentration in the oxidizing ambient. Typically the OED effect decreases as the chlorine concentration increases at least for high temperature oxidations.

Dopant diffusion across material interfaces is modeled in SUPREM with a segregation mechanism. The dopant current, F_s, across an interface from

Process Simulation

material 1 to material 2 is given by

$$F_s = h\left(C_1 - \frac{C_2}{m}\right) \quad (2.27)$$

where m is the segregation coefficient and h is transport coefficient. m and h are specified for each dopant type and each interface type. Both m and h may be activated functions of temperature. The evaporation of dopant from the top of the simulation structure is modeled by setting $C_1 = 0$ and $m = 1$. Chemical predeposition is modeled by setting C_1 equal to either the dopant solid solubility or to any other user specified concentration. For this application, h is set to a large value.

Application Examples

This section describes the simulation of a simplified NMOS process using SUPREM II and SUPREM III. Two vertical slices of the transistor structure in Fig. 2.1 are simulated: the channel region and the source/drain region. Since the field region will not be simulated, the initial process steps related to the island definition and field oxidation may be omitted. The simplified process flow is as follows.

1. Channel Implant: The channel implant consists of two boron implants: 1E12 cm^{-2} at 30 KeV and 5E11 cm^{-2} at 70 KeV. These implants adjust the threshold voltage and control punchthrough. The implants are done through a 40 nm oxide to avoid any damage to the silicon surface and to reduce implant channeling.
2. Gate Oxide: The 40 nm implant oxide is stripped and a 25 nm gate oxide is grown at 900 °C in a dry oxygen ambient. Some of the channel boron is incorporated into the gate oxide as it is grown. This boron is not electrically active and thus does not contribute to the threshold adjustment.
3. Poly Silicon Deposition, Doping and Patterning: Poly silicon is deposited at a relatively low temperature, 650 °C. This temperature cycle will not affect the doping profiles. The poly is doped n^+ using a predeposition process at 950 °C. The bulk silicon under the gate oxide sees this as an inert ambient operation since no oxide is grown on the silicon. The poly is then masked

and etched to form the gate electrode. This etch stops at the gate oxide, which is left in place to protect the silicon during the source/drain implant.
4. Source/Drain Implant: 6e15 cm^{-2} at 80 KeV. This implant is done through the remaining gate oxide. The poly silicon gate protects the channel region from this implant.
5. Source/Drain Reoxidation and Drive-In: A short oxidation cycle at 900 °C in dry oxygen ambient is used to remove any implant damage from the silicon surface. This appears as an inert ambient anneal to the channel region since oxide is grown only in the source/drain region. A final 950 °C, 60 minute anneal activates and drives-in the n^+ source/drain doping profile.

The remaining steps of the process involve dielectric deposition, contact formation, metal deposition and etch, etc. These all involve reasonable low temperatures that will not impact the doping profiles.

Fig. 2.4 shows the SUPREM II and SUPREM III input files for the channel region simulation. Note that the SUPREM II input is usually written in upper case characters. SUPREM III is not case sensitive and any combination of upper and lower case is allowed. In each file the **title** and **comment** lines are for documentation and do not affect the program.

SUPREM II uses the GRID and SUBSTRATE statements (lines 2,3) to define the initial structure. SUPREM III uses the **initialize** statement (line 2). In this case the initial structure is a <100> silicon substrate doped with boron at the 6E14 cm^{-3} level. The thickness of the initial silicon layer is 1 um which is set with the YMAX or **thick** parameters. The grid spacing is set in units of microns with the DYSI or **dx** parameters.

Next the process steps are input in the order that they appear in the actual process. In SUPREM II, each process step is described with a STEP statement. For example, line 5 is a deposition step. In SUPREM II, only oxide may be depositied. The GRTE parameter indicates the deposition rate and the TIME parameter gives the deposition time. SUPREM II automatically places grid points in the deposited layer. Lines 7 and 8 describe the channel implants. The program uses the Pearson IV with exponential tail for these profiles. In line 10, the deposited oxide is etched away. Any dopant that was present in this layer from the channel implants is discarded.

(A)
```
1....TITLE : CHANNEL PROFILE SIMULATION
2....SUBSTRATE ORNT=100,ELEM=B,CONC=6E14
3....GRID DYSI=0.01,YMAX=1.0
4....COMMENT : DEPOSIT 40NM OXIDE
5....STEP TYPE=DEPO,TIME=1,GRTE=0.04
6....COMMENT : IMPLANT THROUGH OXIDE
7....STEP TYPE=IMPL,ELEM=B,DOSE=1E12,AKEV=30
8....STEP TYPE=IMPL,ELEM=B,DOSE=5E11,AKEV=70
9....COMMENT : ETCH OXIDE
10...STEP TYPE=ETCH,TEMP=25
11...COMMENT : GROW GATE OXIDE
12...STEP TYPE=OXID,TEMP=900,TIME=105,MODL=DRY0
13...COMMENT : POLY DOPING
14...STEP TYPE=OXID,TEMP=950,TIME=30,MODL=NIT0
15...COMMENT : SAVE STRUCTURE FOR S/D SIMULATION
16...SAVE FILE=CHNSD,TYPE=B
17...COMMENT : DRIVE IN
18...STEP TYPE=OXID,TEMP=900,TIME=20,MODL=NIT0
19...COMMENT : PRINT AND PLOT RESULTS OF NEXT STEP
20...PRINT HEAD=Y
21...PLOT TOTL=Y,WIND=1.0
22...STEP TYPE=OXID,TEMP=950,TIME=60,MODL=NIT0
23...SAVE FILE=CHNPRO,TYPE=B
24...END
```

(B)
```
1....title : channel profile simulation
2....initialize <100> silicon boron conc=6e14
   + thick=1 dx=0.01
3....comment : deposit 40nm oxide
4....deposit oxide thick=0.04 dx=0.01
5....comment : implant through oxide
6....implant boron dose=1e12 energy=30
7....implant boron dose=5e11 energy=70
8....comment : etch oxide
9....etch oxide all
10...comment : grow gate oxide
11...diffusion temp=900 time=105 dryo2
12...comment : poly doping
13...diffusion temp=950 time=30 nit
14...comment : save structure for S/D simulation
15...save structure file=chnsd
16...comment : drive in
17...diffusion temp=900 time=20 nit
18...diffusion temp=950 time=60 nit
19...comment : print and plot results
20...print layer
21...plot net active
22...save structure file=chnpro
23...stop
```

Fig. 2.4. SUPREM Input Files for the Channel Region Simulation (a) SUPREM II (b) SUPREM III

In SUPREM II, all high temperature anneals and oxidations are designated with the TYPE=OXID parameter; lines 12, 14, 18 and 22. The ambient is designated with the NIT0, DRY0 and WET0 parameters.

After the poly doping operation (line 14) the channel and source/drain regions are treated differently. Therefore the structure is saved (line 16) in a file entitled CHNSD which will be used as the starting point for the source/drain simulation. The TYPE=B parameter specifies a binary file format.

The PRINT statement (line 20) instructs the program to print out some specified information. The HEAD=Y parameter indicates that the "header" should be printed. The "header" includes the thickness and integrated dopant content for the silicon and oxide layers. The PLOT statement (line 21) creates a log-linear plot of the doping profile. TOTL=Y indicates that the net doping level is plotted. The net doping is the absolute value of the n-type concentration minus the p-type concentration. The WIND parameter sets the depth of the plot. Oddly enough, the PRINT and PLOT statements apply to the result of the next process step. Therefore, to obtain the final profile, these two statements must appear before the final anneal cycle, line 22. The last SAVE statement (line 23) stores the final profile for future reference.

SUPREM III uses different statement names for each type of process step. For example, the oxide deposition is input with the **deposit** statement, line 4. The **dx** parameter specifies the grid spacing in the deposited layer. The channel implants are given in lines 6 and 7. The Pearson IV profile is used for these implants. The Pearson IV coefficients were selected to model channeling of the implanted boron and are different from the default parameters in SUPREM III.

In SUPREM III, all high temperature anneals and oxidations are input with the **diffusion** statement, lines 11, 13, 17 and 18. The ambient is designated with the **nit**, **dryo2** and **weto2** parameters.

After the poly doping operation, the simulation structure is saved (line 15) in a binary file entitled **chnsd** which will be used as the starting point for the source/drain simulation. The **structure** parameter specifies the correct file format so that it can be read back into the SUPREM III program. The final results are output with the **print** and **plot** statements, lines 20 and 21. The **layer** parameter instructs the program to print the thicknesses and integrated dopant content of all the layers in the structure. The **plot** statement creates a log-linear plot of the **net active** doping profile. For high doping levels the active concentration will be less than the chemical concentration. The final structure is saved (line 22) for future reference.

Fig. 2.5 shows the input files for the source/drain simulation. The structure created in the channel simulation, file CHNSD, is reloaded into SUPREM II with the LOAD statement, line 5. Then the source/drain implant is input in line 7. A two-side gaussian is used for this implant profile. The reoxidation and final drive-in are described in lines 9 and 14. The final structure is saved in line 15.

The simulation structure is loaded into SUPREM III in the **initialize** statement, line 3. The source/drain implant is specified in line 5. A Pearson IV profile is used for this implant. The reoxidation and final drive-in are input in lines 7 and 9. The final structure is saved in line 13.

The simulated doping profiles for the channel and source/drain regions are presented in Fig. 2.6. The differences between the predictions of SUPREM II and SUPREM III are due to the different implant profiles and the somewhat different default diffusion constants. In general the program user needs to adjust some of the model parameters to provide agreement with experimental observations. Chapter 6 goes into this issue in more detail.

IN SUPREM II the MODL statement is used to modify the model coefficients. For example, one could modify the parameters related to dry oxidation and designate the new coefficient set as model DRY1. This could then be input in a oxidation operation with the MODL=DRY1 parameter. In SUPREM II there is no way to save the new coefficient set for use in future simulations. Therefore any coefficient modifications must be repeated in each input file.

SUPREM III offers several statement types that can be used to modify the model coefficients. For example, the models related to an impurity such as boron may be modified in a **boron** statement. The model coefficients related to the silicon may be modified with a **silicon** statement. The segregation coefficients may be modified with a **segregate** statement. The modified coefficient set can be saved in a file and used in future simulations by specifying the coefficient file in the **initialize** statement.

(A)
```
1....TITLE : SOURCE/DRAIN SIMULATION
2....SUBSTRATE ORNT=100,ELEM=B,CONC=6E14
3....GRID DYSI=0.01,YMAX=1.0
4....COMMENT : LOAD STRUCTURE FROM CHANNEL SIMULATION
5....LOAD FILE=CHNSD,TYPE=B
6....COMMENT : IMPLANT SOURCE/DRAIN
7....STEP TYPE=IMPL,ELEM=AS,DOSE=6E15,AKEV=80
8....COMMENT : REOXIDATION
9....STEP TYPE=OXID,TEMP=900,TIME=20,MODL=DRY0
10...COMMENT : PRINT AND PLOT RESULTS OF LAST STEP
11...PRINT HEAD=Y
12...PLOT TOTL=Y,WIND=1.0
13...COMMENT : FINAL DRIVE IN
14...STEP TYPE=OXID,TEMP=950,TIME=60,MODL=NIT0
15...SAVE FILE=SDPRO,TYPE=B
16...END
```

(B)
```
1....title : source/drain profile simulation
2....comment : load structure from channel simulation
3....initialize structure=chnsd
4....comment : source/drain implant
5....implant arsenic dose=6e15 energy=80
6... comment : reoxidation
7....diffusion temp=900 time=20 dryo2
8....comment : final drive in
9....diffusion temp=950 time=60 nit
10...comment : print and plot results
11...print layer
12...plot active net
13...save structure file=sdpro
14...stop
```

Fig. 2.5. SUPREM Input Files for the Source/Drain Region Simulation (a) SUPREM II (b) SUPREM III

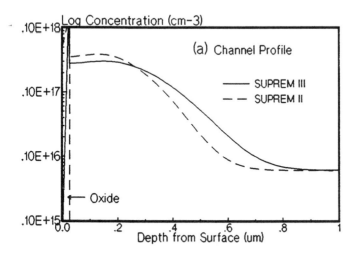

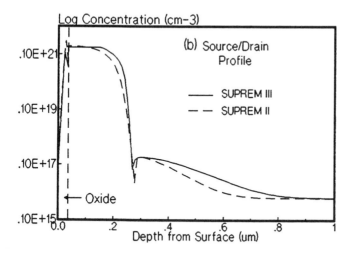

Fig. 2.6. Simulated Doping Profiles (a) Channel Region (b) Source/Drain Region

2.3 SUPRA : 2-D Process Simulator

SUPRA simulates the incorporation and redistribution of impurities in a two-dimensional (2-D) cross-section of a device as indicated in Fig. 2.1. Such two-dimensional structures are created by modeling actual lithographic patterning. A photoresist layer deposition in Fig. 2.7, for example, is formed by an ETCH card. Four parameters (START, CORNER, END and THICK) give enough freedom to create arbitrary etch profiles. However, re-entrant angles resulting from undercutting are not allowed because such a shape is difficult to express numerically with an array that represents a single value of a layer thickness at a horizontal position. Impurity concentrations are calculated only within the thermal oxide and substrate regions. The mask layers are used only as barriers against ion implantation and surface diffusion.

Input commands and the internal organization of SUPRA are similar to those of SUPREM II. Specification of the two-dimensional device structures, however, requires the use of a coordinate system. The horizontal coordinate is defined as x and corresponds to variation parallel to the bottom surface of the device. The vertical coordinate is defined as y and corresponds to variation normal to the bottom surface of the device. The origin of coordinates is defined as the leftmost point of the original semiconductor/oxide interface, with positive x to the right and positive y downward into the substrate.

Finite-difference grids in two dimensions are easily generated by using X.GRID and Y.GRID statements for the x and y directions. Nonuniform grids are automatically generated, depending upon the number of nodes and spaces specified in the x and y directions. This allows fast computation as well as high resolution for regions where concentrations change rapidly. More details on the grid generation will be discussed in a following example section.

Two-Dimensional Process Models

Process models of the SUPRA program consist of two main parts. One is an analytical part containing closed forms of analytical equations, from which

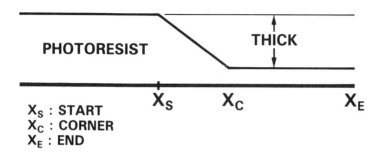

Fig. 2.7. Structure definition in SUPRA. START, CORNER, and END represent the starting, corner, and ending positions. THICK is the thickness of the layer to be etched.

solutions are readily obtained at any position without numerical iterations. The other is a numerical part in charge of specific calculations for solving the nonlinear diffusion equation. Analytical solutions are used to model changes in the overall device structure (i.e. deposition/etch and oxide growth), ion implantation, and boron diffusion. This analytic technique is fast in computation and flexible in grid allocation. It is, however, limited to low-concentration diffusion. Inert drive-in involving high impurity concentration of arsenic-implanted source/drain regions is treated by the numerical solution.

a) Ion Implantation

The spatial distribution of implanted atoms is assumed to be a Gaussian function in the vertical and lateral direction when implanted through an infinitesimal mask window (actually a point source)[2.20]. When this Gaussian function is integrated for a finite window with a uniform thickness as shown in Fig. 2.8 from x_k to x_{k+1}, the profile is given below as

$$I_k(x, y) = \frac{I_{max}}{2} \exp\left[-\frac{(y-R_p-d_k)^2}{2\Delta R_p^2}\right]\left[-\text{erfc}\left(\frac{x-x_k}{\sqrt{2}\Delta x}\right) + \text{erfc}\left(\frac{x-x_{k+1}}{\sqrt{2}\Delta x}\right)\right] \quad (2.28)$$

where I_{max} is the peak concentration and erfc(x) is the complementary error function of x. The parameters R_p, ΔR_p and Δx are the projected range, and vertical and lateral standard deviations respectively. It should be noted that the

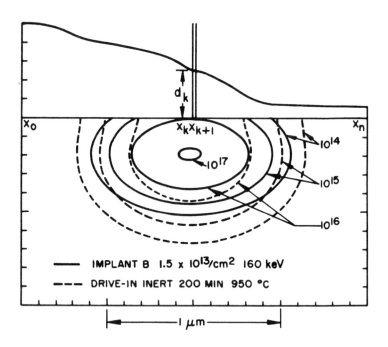

Fig. 2.8. Impurity distributions implanted through a small mask window after implantation (solid lines) and after inert drive-in (broken lines).

profile implanted through the window with a finite width becomes a Gaussian distribution multiplied by a complementary-error function in the lateral direction. The impurity profile, when implanted through an arbitrary mask as shown in Fig. 2.8, can be expressed by a summation of the profiles through all mask segments as

$$I(x, y) = \sum_{k=1}^{n} I_k(x, y) \qquad (2.29)$$

where k represents the kth mask segment.

b) *Low Concentration Inert Drive-in*

If the impurity concentration is lower than the intrinsic carrier concentration n_i, as in the boron channel-stop region, the diffusivity can be assumed constant. The diffusion Eq. (2.14) becomes linear when the drift term is ignored. The same superposition technique used for implantation can be applied to this inert drive-in condition. The inert drive-in solution for the profile described in Eq. (2.28) is given by

$$N_i(x, y, t) = N_{ix}(x, t) N_{iy}(y, t) \tag{2.30a}$$

$$N_{ix}(x, t) = -\frac{1}{2}\text{erfc}\left(\frac{x - x_k}{\sqrt{2\Delta x^2 + 4Dt}}\right) + \frac{1}{2}\text{erfc}\left(\frac{x - x_{k+1}}{\sqrt{2\Delta x^2 + 4Dt}}\right) \tag{2.30b}$$

$$N_{iy}(y, t) = \frac{I_{max} \Delta R_p}{2\sqrt{2Dt + \Delta R_p^2}} [\Omega(y, t) + \Omega(-y, t)] \tag{2.30c}$$

$$\Omega(y, t) = \exp\left[-\frac{(y - R_p - d_k)^2}{4Dt + 2\Delta R_p^2}\right] \left\{ 2 - \text{erfc}\left[\frac{y \Delta R_p^2 + 2(R_p - d_k)Dt}{\Delta R_p \sqrt{2Dt(4Dt + 2\Delta R_p^2)}}\right] \right\} \tag{2.30d}$$

where the subscript i refers to inert ambient conditions. D is the impurity diffusivity, t is the time of the diffusion step, d_k is the effective silicon thickness of the layers above the substrate through which the ion implantation is performed. The profile resulting from an inert drive-in of the implant through one mask segment is shown in Fig. 2.8 (broken lines). Once again, the complete drive-in profile is constructed by superposing the solution for each mask segment.

c) *Moving Boundary Diffusion*

The one-dimensional profile following oxidation is approximated by adding a correction factor to the inert drive-in concentration [2.21] as

$$N_o(y, t) = N_i(y, t) + N_c(y, t) \tag{2.31}$$

where the correction factor $N_c(y,t)$ is expressed by a term involving a complementary error function. In this approach the correction function basically subtracts an appropriate fraction of dopant to account for segregation and out-diffusion into the oxide. A boundary condition which properly determines the correction term has been found, and extensive analytical and numerical results

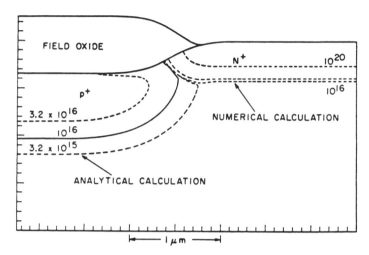

Fig. 2.9. Cross-section of NMOSFET near the source and the channel-stop region simulated by SUPRA.

have been presented elsewhere [2.22]. For low concentration diffusion, this analytic method gives agreement typically better than 10% with numerical calculations, and it has a computation-time advantage of more than a factor of ten [2.23].

A quasi two-dimensional profile has been obtained by assuming that the oxide layer grows only in the vertical direction during a semi-recessed oxidation [2.22]. However, the effect of lateral oxide growth may not be negligible in a case such as fully-recessed oxidation. A more generalized boundary condition is given below as

$$D\nabla N_o(x, y, t)\vert_{x_f, y_f} = A_o N_o(x_f, y_f, t) g(x_f, y_f, t) \mathbf{n}(x_f, y_f, t) \quad (2.32a)$$

$$A_o = \frac{1}{m} - \alpha \quad (2.32b)$$

where the coefficient m is the equilibrium segregation factor and α is the volumetric ratio of silicon consumed in forming one unit of oxide (0.44). The quantity $g(x_f, y_f, t)$ is the oxide growth rate, $\mathbf{n}(x_f, y_f)$ is a unit vector normal to the SiO_2 interface, and (x_f, y_f) is the final interface point closest to (x, y) as indicated in Fig. 2.9. Provided that $N_o(x, y, t)$ can be separated in the x and y variables, the above boundary condition is also separable in the vertical and lateral directions. The resulting 2-D profile is given by

Process Simulation

$$N_o(x, y, t) = N_{ox}(x, t) N_{oy}(y, t) \tag{2.33a}$$

$$N_{ox}(x, t) = N_{ix}(x, t)\left[1 + A_x \text{erfc}\left(\frac{x - x_f}{2\sqrt{Dt_{ox}}}\right)\right] \tag{2.33b}$$

$$N_{oy}(y, t) = N_{iy}(y, t)\left[1 + A_y \text{erfc}\left(\frac{y - y_f}{2\sqrt{Dt_{ox}}}\right)\right] \tag{2.33c}$$

where the subscripts o and i indicate oxidizing and inert conditions, respectively. The Dt terms implicit in N_{ix} and N_{iy} represent the diffusivity-time product for all high temperature steps including inert drive-in. The correction terms involve only the moving boundary and segregation effects, hence Dt_{ox} represents the diffusivity-time product for oxidation steps only.

The coefficient A_x and A_y are obtained from Eq. (2.33) and the boundary condition Eq. (2.32) as

$$A_z = \frac{A_o g_z N_{iz}(z_f, t) - D\dfrac{\partial N_{iz}(z, t)}{\partial z}\bigg|_{z=z_f}}{(A_o g_z + D/\sqrt{\pi Dt_{ox}}) N_{iz}(z_f, t) - D\dfrac{\partial N_{iz}(z, t)}{\partial z}\bigg|_{z=z_f}} \tag{2.34}$$

where z is either the x or y variable. Again g_z is the oxide growth rate in either the vertical or lateral direction. The boron profile obtained by this analytical technique at the channel-stop region is shown in Fig. 2.9. The two-dimensional oxide shape resulted from LOCOS is approximated by empirically determined functions. More rigorous treatment of non-uniform oxidation is the topic of the next section.

d) High-Concentration Diffusion

If the impurity concentration is sufficiently large compared to the intrinsic electron concentration at the diffusion temperature, then diffusivity will no longer be constant throughout the simulation space. The diffusion equation then becomes nonlinear and must be solved numerically. Also, there may be several impurities present simultaneously which can interact by way of electric field or Fermi level effects. Also impurity clustering can become important.

The diffusion equation for each impurity species is obtained by extending Eq. (2.14) to two dimensions as

$$\frac{\partial C}{\partial t} = -\nabla \cdot \mathbf{J} \tag{2.35a}$$

$$\mathbf{J} = -D\nabla N \pm \frac{-q}{kT}(DN\nabla\phi) \tag{2.35b}$$

$$\phi = \frac{kT}{q}\ln\left(\frac{n}{n_i}\right) \tag{2.35c}$$

where the basic models on the diffusivity D and the total/active concentrations are the same as SUPREM II discussed in section 2.2. The potential ϕ can be eliminated by setting the net carrier concentration $(n-p)$ equal to the net active impurity concentration (U) and then using the relationship $np = n_i^2$.

$$n = \frac{U + \sqrt{U^2 + 4n_i^2}}{2} \tag{2.36}$$

Assuming that there is a differentiable relation between the electrically active concentration N and the total chemical concentration C, the expression for flux can be written as

$$\mathbf{J} = -D_{\it eff}\nabla C \pm \frac{-DN}{\sqrt{U^2 + 4n_i^2}}\nabla U \tag{2.37a}$$

$$D_{\it eff} \equiv D\frac{dN}{dC} \tag{2.37b}$$

$D_{\it eff}$ is often referred to as the effective diffusivity and will in general be less than the true diffusivity D since only a portion of the total impurity concentration is mobile at high concentration due to clustering. The numerical solution of the diffusion equation is formulated by approximating the continuous concentration profile with its values on a network of nodes within the device boundaries. The grid structure chosen for this numerical simulation is rectangular with nonuniform spacing in both spatial directions. The finite-difference approximation of Eqs. (2.35) at a node surrounded by its four nearest neighbors [2.23] is

Process Simulation

$$\frac{\partial C_o}{\partial t} = \sum_{m=1}^{4}\left[B_m(C_m - C_o) \pm E_m(U_m - U_o)\right] \quad (2.38)$$

where B_m and E_m are coefficients related to the grid spaces and the electron concentration [2.24].

For approximating the time derivative $\partial C_o/\partial t$, it has been found that a three-level time discretization scheme is advantageous for treating nonlinear parabolic equations. As an approximation,

$$\frac{\partial C_o}{\partial t} \approx \frac{C_o^{n+1} - C_o^{n-1}}{k} \quad (2.39)$$

where k is the value of the time step and the superscripts $n+1$ and $n-1$ denote values at the future and previous time levels, respectively.

NMOS Transistor Simulation

This section presents an example of a complete NMOS transistor fabrication process. The field region oxide growth and boron redistribution are simulated analytically, making the assumption that the boron concentration in the field region is low enough that its diffusivity will be constant. The program is then switched to numerical mode where the arsenic source/drain regions are implanted and driven in. In numerical mode the arsenic clustering and impurity interaction are taken into account. The structure file is saved and is used both for two-dimensional plots and to create the complete transistor structure.

a) Analytic Simulation of a Fully-recessed Oxide Isolation Region

The input command file shown in Fig. 2.10 simulates the initial portion of an NMOS process through the definition of the polysilicon gate. The substrate is p-type with <100> orientation and 20 Ω-cm resistivity. The structure is 5 µm deep and 3 µm wide. The height is specified to be 1 µm which allocates space for the thermal oxide that will be grown. The horizontal grid has 0.1 µm spaces at the device edges, decreasing toward the center of the device. Nodes inside the region are automatically generated with a nonuniform but smooth (actually quadratic) distribution.

The program is started out in analytic mode, since only in analytic mode can the silicon substrate be etched or thermal oxide be grown. After the silicon substrate is etched, an 80 nm pad oxide, a 70 nm nitride layer, and a 2 μm photoresist layer are deposited and selectively etched. The field region is implanted with boron and subsequently oxidized for 3 hours at 1000 °C in wet oxygen. The nitride is then stripped and an unmasked enhancement implant is performed. The polysilicon gate is deposited and etched with sloping sides. After switching to the numerical mode, the current result is saved into a file by SAVE card for further simulation. The current result is plotted two-dimensionally by using a PLOT.2D card. Equi-concentration lines are plotted within the range specified by FIRST and LAST. The next SUPRA input example file reads the saved result and plots the structure (Fig. 2.11).

b) Numerical Source/Drain Simulation

The input command file shown in Fig. 2.12 performs the numerical source/drain simulation. The initial structure is defined by loading the structure file EX2AST. This will be a high concentration source/drain simulation, so the program is placed in numerical mode. Arsenic is implanted, using the field oxide and polysilicon gate to define where the arsenic enters the silicon. The drive-in is then performed for 30 minutes at 1000 °C in an inert ambient and the resulting structure saved in the structure file EX2NST. The discretized boundaries of the thermal oxide and silicon substrate used in the numerical solution are plotted in Fig. 2.12. The contours of constant arsenic concentration are plotted with a dashed line type to distinguish them from the boron contour. Note that the electric field generated by the arsenic tends to pull the boron into the source/drain region where it segregates into the thermal oxide.

The whole transistor structure can be generated when the output file EX2NST is read twice. The second LOAD card, however, needs a parameter REFLECT to convert the structure of Fig. 2.12 to a mirror image. SUPRA is also capable of extending or shrinking the right or left side of the original structure during loading.

Process Simulation

```
 1... COMMENT    Example 2 - NMOS transistor simulation
 2... COMMENT    Analytic recessed field region simulation

 3... STRUCTURE  P-TYPE ORIENTATION=100 DEPTH=5 WIDTH=3 HEIGHT=1
  ...  +         RESISTIVITY=20
 4... X.GRID     H1=.1   H2=.1   WIDTH=2   N.SPACES=34
 5... X.GRID     H1=.1   H2=.1   WIDTH=1   N.SPACES=10
 6... Y.GRID     H1=.1   H2=.02  DEPTH=1   N.SPACES=15
 7... Y.GRID     H1=.02  H2=.2   DEPTH=5   N.SPACES=40
 8... END        Structure definition

 9... COMMENT    Start out in analytic mode
10... ANALYTIC

11... COMMENT    Silicon etch to recess field regions
12... ETCH       SILICON  THICK=.4  START=.7  END=0  ANGLE=54.7

13... COMMENT    Pad oxide deposition
14... DEPOSIT    OXIDE  THICKNESS=.08

15... COMMENT    Nitride deposition and field region mask
16... DEPOSIT    NITRIDE  THICKNESS=.07
17... DEPOSIT    PHOTO.RESIST  THICKNESS=2
18... ETCH       PHOTO.RESIST  START=.7  END=0  ANGLE=90
19... ETCH       NITRIDE  START=.7  END=0  ANGLE=90

20... COMMENT    Boron field implant
21... IMPLANT    BORON  DOSE=5E12  ENERGY=100

22... COMMENT    Strip photoresist
23... ETCH       PHOTO.RESIST  START=0  CORNER=0  END=3

24... COMMENT    Field oxidation
25... OXIDIZE    TIME=180  TEMPERATURE=1000  WET

26... COMMENT    Strip nitride
27... ETCH       NITRIDE  START=0  CORNER=0  END=3

28... COMMENT    Unmasked enhancement implant
29... IMPLANT    BORON  DOSE=1E11  ENERGY=75

30... COMMENT    Poly silicon deposition and gate mask
31... DEPOSIT    POLYSILICON  THICKNESS=.5
32... ETCH       POLYSILICON  START=2.3  CORNER=2.1  END=0

33... COMMENT    Switch into numerical mode and save the structure
34... NUMERICAL
35... SAVE       STRUCTURE=EX2AST::-25
```

Fig. 2.10. SUPRA Input Simulating Recessed-oxide Isolation

```
1... TITLE       Example 2 - NMOS transistor simulation
2... COMMENT     Two-dimensional analytic simulation plot

3... COMMENT     Load analytic solution
4... STRUCTURE
5... LOAD        STRUCTURE=EX2AST::-25
6... END         Structure definition

7... PLOT.2D     X.MAX=5  Y.MAX=2
8... BOUNDARY    SMOOTH
9... CONTOUR     BORON  FIRST=1E15  LAST=1E17  RATIO=10  OXIDE
10... END        2-D plot

*** END SUPRA ***
```

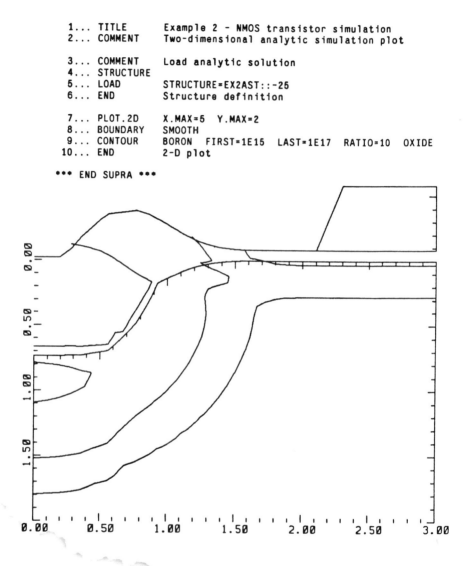

Fig. 2.11. SUPRA Output Showing a Two-dimensional Boron Distribution near the Channel-stop Region

Process Simulation

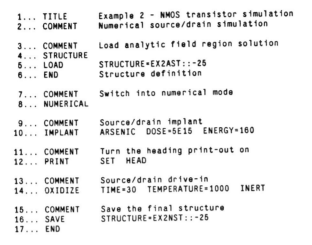

```
 1... TITLE       Example 2 - NMOS transistor simulation
 2... COMMENT     Numerical source/drain simulation

 3... COMMENT     Load analytic field region solution
 4... STRUCTURE
 5... LOAD        STRUCTURE=EX2AST::-25
 6... END         Structure definition

 7... COMMENT     Switch into numerical mode
 8... NUMERICAL

 9... COMMENT     Source/drain implant
10... IMPLANT     ARSENIC  DOSE=5E15  ENERGY=160

11... COMMENT     Turn the heading print-out on
12... PRINT       SET  HEAD

13... COMMENT     Source/drain drive-in
14... OXIDIZE     TIME=30  TEMPERATURE=1000  INERT

15... COMMENT     Save the final structure
16... SAVE        STRUCTURE=EX2NST::-25
17... END
```

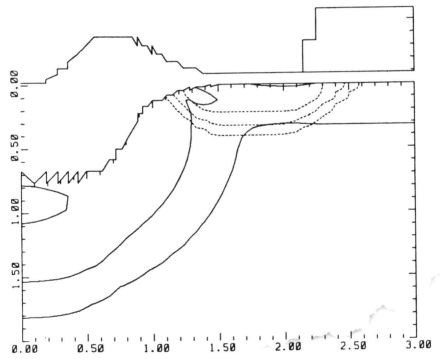

Fig. 2.12. SUPRA Input and Output Simulating the Source/Drain Region

2.4 SOAP : 2-D Oxidation Simulator

In today's VLSI technology, such extrinsic materials as polysilicon, nitride and oxide are deposited and patterned. The silicon substrate itself is etched to form a recessed surface. As a consequence, the silicon surface very often becomes uneven when a thermal oxidation process starts. The shapes of the oxide grown on such nonplanar surfaces are surprisingly different from those expected on a flat surface. For instance, the oxide on an etched silicon substrate forms a cusp that becomes thinner at an inside corner as shown in Fig. 2.13 [2.25]. At the outside silicon corner, the oxide thickness is also smaller than at the sidewalls. A similar phenomenon occurs when a gate oxide is grown in the vicinity of a thick field oxide edge [2.26]. This oxide thinning effect often causes a device failure because the breakdown voltage becomes lower near the corners. LOCOS is an excellent isolation technique that allows a self-aligned channel stop implantation and causes small parasitic capacitance. The isolation region typically takes almost half of the chip area and the transition region from the thin pad oxide under the nitride to the thick field oxide (so called bird's beak) becomes a non-negligible factor in a sub-micron transistor. Reduction of the bird's beak becomes one of the key issues in improving packing densities and device performance (see Chapter 11).

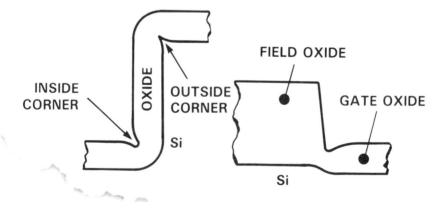

Fig. 2.13. Examples of non-planar oxidation.

Two-Dimensional Oxidation Model

The two-dimensional oxidation model introduced in 1982 [2.27] is based on the physical behavior that the oxide can flow at high temperature. The viscous flow phenomenon was first observed in an experiment that a silicon wafer with thermal oxide on one side is heated in a nonoxidizing condition [2.28,2.29]. Since the thermal expansion of oxide is larger than silicon, the wafer bends toward the silicon substrate when the temperature is lower than 960 °C. However, the wafer remains flat above this critical temperature, which suggests that stress due to the thermal mismatch is relaxed through viscous flow of oxide. The viscosity of oxide has been measured from the curvature of wafers as

$$\mu(T) = \mu_o \exp(E_\mu / kT) \quad (2.40)$$

where μ_o = 1.586E10 poises, and E_μ = 5.761 eV. This viscous flow is a dominant mechanism for the relaxation of stress induced by volume expansion during thermal oxidation. It should be noted that the viscosity increases tremendously as temperature goes down. Stress, therefore, may be insufficiently relaxed at temperatures below the glass-transition temperature (960 °C) causing defect generation. The two-dimensional oxidation model in SOAP is based on viscous flow [2.30].

As discussed in section 2.2, the oxidant diffuses in steady state. The flux conservation of Eq. (2.4) and Eq. (2.5) is replaced by a generalized diffusion equation as

$$D_{eff} \nabla^2 C = \frac{\partial C}{\partial t} \approx 0 \quad (2.41)$$

where the effective diffusivity D_{eff} is assumed to be independent of oxidant concentration C or stress as in the linear-parabolic model. Boundary conditions for Eq. (2.41) are also obtained from the flux conditions as

$$\mathbf{F} \cdot \mathbf{n}(x, y) = \begin{cases} kC(x, y) & \text{on } S1 \\ -h[C_x - C(x, y)] & \text{on } S2 \\ 0 & \text{on } S3, S4, S5 \end{cases} \quad (2.42)$$

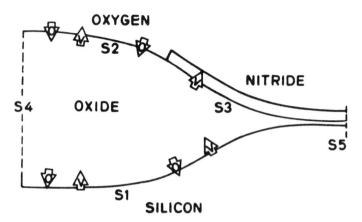

Fig. 2.14. Boundary conditions of the 2-D oxidation.

where **F** is the oxidant flux and **n** is a unit vector normal to the oxide surface (positive in the outward direction from the oxide bulk). The boundaries S1, S2 and S3 are the oxide/silicon interface, free-oxide surface, and nitride boundary, respectively, as defined in Fig. 2.14. C_x is the oxygen concentration in the gaseous ambient. These non-homogeneous conditions cause the oxidant distribution at the silicon/oxide interface to be non-uniform when a nitride layer is involved as in Eqs. (2.42) or the normal vector **n** changes direction. The oxidation rate, as a consequence, is different from place to place at the interface. The volume expansion rate, equivalent to the velocity of an oxide element close to the interface, can be given as

$$\mathbf{V}(x, y, t) = -(1-\alpha)\frac{F(x, y, t)}{N_1}\mathbf{n}(x, y, t) \quad \text{on } S1 \qquad (2.43)$$

where the minus sign indicates that the velocity is toward the oxide bulk and α is the ratio of the consumed silicon volume to the grown oxide volume. N_1 is the number of oxidant molecules per unit volume of the oxide. Every element in the already existing oxide bulk moves because of the newly grown oxide layer. The velocity, however, is very slow because the oxidation rate is at most a few angstrom per second even in a steam ambient.

Due to the unique slow-viscous flow behavior of oxidation, the motion of oxide elements can be described by a simplified Navier-Stoke's equation as

Process Simulation

$$\mu\nabla^2 \mathbf{V} = \nabla P \qquad (2.44)$$

where μ is the oxide viscosity, $\mathbf{V}$ is the velocity, and P is the pressure. This first equation indicates that the viscous damping force is balanced by the pressure gradient. Since the compressibility of oxide is very small (2.7×10^{-12} cm^2/dyne), the oxide is assumed to be incompressible. This incompressibility characteristics requires that the density remain constant with respect to time. The continuity equation, consequently, is approximated below as

$$-\frac{d\rho}{dt} = \rho \nabla \cdot \mathbf{V} \approx 0 \qquad (2.45)$$

where ρ is the oxide density. The velocity obtained from the O$_2$ concentration in Eq. (2.42) serves as a boundary condition on S1. On the oxide surfaces, boundary conditions are derived from an assumption that the internal pressure is balanced with external stress. Namely P becomes the ambient pressure on S2 and the nitride stress on S3, added by surface tension.

In the two-dimensional oxidation system, the ultimate solution we are looking for is the oxide boundary position that moves continuously. This moving-boundary problem is extremely difficult to solve with a conventional numerical method based on finite-difference or finite-element. The SOAP program adopted a different approach known as a boundary-value technique in which grid points are located on the boundaries and an integral form of Eqs. (2.41) and Eqs. (2.44) is solved by using Green's function. Initially SOAP takes an initial guess that the pressure distribution is constant and thus $\nabla P = 0$. The velocity field calculated from this guessed pressure distribution would not satisfy the zero divergence of Eq. (2.45) everywhere. The resulting $\nabla \cdot \mathbf{V}$ is fed back to obtain a new pressure distribution and then the velocity field is again calculated from the new pressure. This velocity/pressure iteration scheme continues until the incompressibility condition is satisfied within an allowable range. The flow chart of this algorithm is given in Fig. 2.15. Details of the numerical method as well as the physical model are discussed elsewhere [2.30].

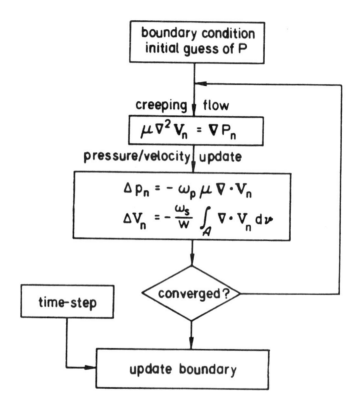

Fig. 2.15. Flow Chart of the Velocity-pressure Iteration Method

Process Simulation

Application Examples

SOAP is applied to a local oxidation in which 600 nm field oxide is grown from a 10 nm pad oxide with a 100 nm nitride layer. The input file for this example is shown in Fig. 2.16. The starting oxide shape is defined by several GRID cards placed between STRUCTURE and END cards. A GRID card allocates node points on a straight segment of the initial oxide surface. The first GRID card assumes that the starting position of the first segment is (0,0). The end position of the segment is given by the two parameters (X.POSIT, Y.POSIT). The distances between two adjacent nodes are determined by the total number of spaces within the segment and the first/last spaces in the same way as in SUPRA. However, to give sufficient flexibility to the segment orientation, two different notations are allowed for the first and last spaces. Either HX1 or HY1, for an example, can be chosen for a segment that is neither exactly in the horizontal nor in the vertical direction. If the segment is only in the horizontal direction, HX1 and HX2 must be used because there is no y-direction change. The boundary condition that the oxide surface is facing is also specified in the GRID card. The present version of SOAP accepts SILICON, NITRIDE, OXYGEN, or REFLECT boundary condition which specifies that the contacting material is a silicon substrate, a nitride layer, oxygen ambient, or oxide bulk with infinite extension, respectively. One should be cautioned that the position specified by the last GRID card is connected to the starting point (0,0) with a reflecting boundary condition.

The following NUMERICAL card is to control the numerical convergence, simulation time step, and orientation effect. The parameter DOX = 0.05 indicates that the incremental oxide thickness between two simulation steps is fixed by 0.05 μm and therefore the time step is determined by the oxide thickness. When NO.PRESS is specified in the NUMERICAL card, the program is forced to iterate only once in the pressure-velocity iteration algorithm. Although this mode gives an approximate solution that the pressure is constant, fairly close oxide shapes can be obtained with much shorter computation time. Therefore, it is recommended to use this NO.PRESS mode to check the current input file including the definition of the initial structure as well as the final oxide shape expected. The PHYSICAL card allows users to

```
Comment    Input file SOAP EX2 - Semi-Recessed Oxidation with Pressure
$                             Equilibration
Comment    Delimit lines of title with slashes; '*' => 'degree'
TITLE      /600 nm Semi-ROX/100 nm nitride/10 nm pad/950*C

Comment    Save data in file SOAP EX2D0 (0 is default)
SAVE

Comment    Specify x and y plot bounds; display outline, title,
$          all contours, first and last nitride edge and nodes
PLOT.2D    X.MIN=0  X.MAX=2.0 Y.MIN=-0.5 Y.MAX=0.6 OUTLINE TITLE
+          CONT.ALL NIT.1ST NIT.LAST NOD.1ST NOD.LAST

Comment    Define structure (STRUCTURE and END required)
STRUCTURE
GRID       SILICON  X.POSI=2.000  Y.POSI=0.000  N.SP=20 HX1=.2  HX2=.2
GRID       REFLECT  X.POSI=2.000  Y.POSI=0.010  N.SP=1  HY1=.01 HY2=.01
GRID       OXYGEN   X.POSI=1.000  Y.POSI=0.010  N.SP=10 HX1=.2  HX2=.01
GRID       NITRIDE  X.POSI=0.000  Y.POSI=0.010  N.SP=10 HX1=.01 HX2=.2
+          THICK=0.15
END

Comment    Parameters for nitride plastic-elastic stress model; turn on
$          stress-dependent diffusion and reactivity
PHYSICAL   nit.len=.7  plastic=0.02 strdif.on strrea.on

Comment    Equilibrate pressure at each timestep (max 20 iterations);
$          set orientation effect on, set delta oxide = .05 nm
NUMERICAL  MAX=20 PRESS.CALC ORIENT DOX=.05

Comment    Oxidation parameters
OXIDIZE    TEMPERAT=1000  FOXIDE=0.6  IOXIDE=.010 WET BB.DELTA=.001


Comment    Input file SOAP EX2:1 - LOAD Output from Example 2
$          and Plot Stress Along Silicon Surface

Comment    Delimit lines of title with slashes; '*' => 'degree'
TITLE      /600 nm Semi-ROX/100 nm nitride/10 nm pad/950*C

Comment    Select type of 1D plot; lower (silicon) surface; specify plot
$          bounds; y axis spacing; specify outline,title,plot every other
PLOT.1D    STRESS LOWER  LEFT=0.0 RIGHT=2.0 BOTTOM=-2.0E09 TOP=3.2E09
+          YMAJ.SP= 1.0E09 YMIN.SP=0.1E09 OUTLINE TITLE DELTA.IN=2

Comment    LOAD file SOAP EX2D0 (SAVEd in SOAP EX2); no continuation
LOAD       no.cont
```

Fig. 2.16. SOAP Input File Simulating a Semi-recessed Oxide

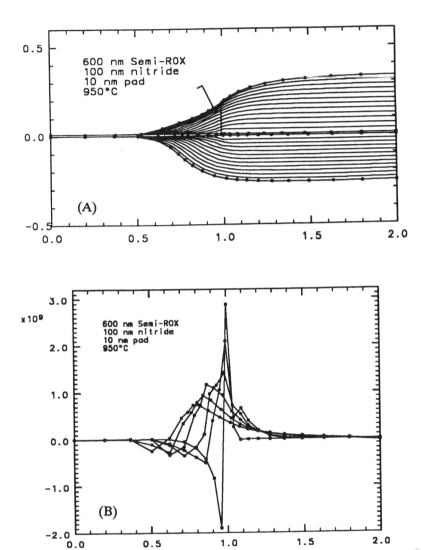

Fig. 2.17. SOAP Output (a) Semi-recessed Oxide Shape (b) Stress Distribution at the Oxide/Silicon Interface

be able to change such parameters related to their own process conditions as oxide growth rate constants. If a user wants to see intermediate oxide shapes during calculation, he can specify a PLOT.2D card beforehand. The OXIDIZE card is to initiate an oxidation step. The final oxide thickness or the total oxidation time can be specified. The initial oxide thickness should be consistent with the initial oxide structure defined by GRID cards. The simulation result is shown in Fig. 2.17.

References

[2.1] D.A. Antoniadis, S. E. Hansen, R. W. Dutton, and A. G. Gonzales, "SUPREM I - A Program for IC Process Modeling and Simulation," SEL 77-006, Stanford Electronics Laboratories, Stanford University, Calif., May 1977.

[2.2] D.A. Antoniadis, S. E. Hansen, and R. W. Dutton, "SUPREM II - A Program for IC Process Modeling and Simulation," TR 5019.2, Stanford Electronics Laboratories, Stanford University, Calif., June 1978.

[2.3] C.P. Ho and S. E. Hansen, "SUPREM III - A Program for IC Process Modeling and Simulation," TR SEL 83-001, Stanford Electronics Laboratories, Stanford University, Calif., July 1983.

[2.4] D. Chin, M.R. Kump, and R.W. Dutton, "SUPRA : Stanford University PRocess Analysis Program," Stanford University Laboratories, Stanford University, Stanford, Calif., July 1981.

[2.5] D. Chin, M.R. Kump, H.G. Lee, and R.W. Dutton, "Process Design Using Two-Dimensional Process and Device Simulators," *IEEE Trans. on Electron Devices* **ED-29**, Feb. 1982, pp. 336-340.

[2.6] D. Chin and R.W. Dutton, "SOAP : Stanford Oxidation Analysis Program," Stanford University Laboratories, TR SEL 83-002, Stanford University, Stanford, Calif. Aug. 1983.

[2.7] B.R. Penumalli, "A Comprehensive Two-Dimensional VLSI Process Simulation Program - BICEPS", *IEEE Trans. on Electron Devices* **ED-36**, Sept 1983, pp. 986-992.

[2.8] G.E. Smith, III, and A.J. Steckl, "RECIPE - A Two-Dimensional VLSI Modeling Program," *IEEE Trans. on Electron Devices* **ED-29**, Feb 1982, pp. 216-221.

[2.9] K.A. Salsburg, and H.H. Hensen, "FEDSS - Finite-Element Diffusion - Simulation System," *IEEE Trans. on Electron Devices* **ED-30**, Sept 1983, pp. 1004-1011.

[2.10] J. Linhard, M. Scharff, and H. Schiott, "Range Concepts and Heavy Ion Ranges," *K. Dan. Vidensk. Selsk., Mat. Fys. Medd.,* **33**(14), 1963

[2.11] T.M. Liu and W.G. Oldham, "Channeling Effect of Low Energy Boron Implant in <100> Silicon," *IEEE Electron Device Letters,* **EDL-4**, Mar. 1983, pp. 59-62.

[2.12] W.K. Hofker, "Implantation of Boron In Silicon," *Philips Research Reports Supplements,* no. 8, 1975.

[2.13] M.G. Kendall and A. Stuart, *The Advanced Theory of Statistics,* (Charles Griffin, London, 1958).

[2.14] H. Ryssel, G. Prinke, K. Haberger and K. Hoffmann, "Range Parameters of Boron Implanted into Silicon," *Appl. Phys.,* **24**, Jan 1981, pp. 39-43.

[2.15] M. Simard-Normandin and C.Slaby, "Empirical Modeling of Low Energy Boron Implants in Silicon," *J. Electrochem. Soc.,* **132**, Sept 1985, pp. 2218-2223.

[2.16] B.E. Deal and A.S. Grove, "General Relationship for the Thermal Oxidation of Silicon," *J. Appl. Phys,* **36**(12), Dec 1965, pp. 3770-3778.

[2.17] C.P. Ho, J.D. Plummer, B.E. Deal, and J.D. Meindl, "Thermal Oxidation of Heavily Phosphorus Doped Silicon," *J. Electrochem. Soc.*, **125**, Apr 1978, pp. 665-671.

[2.18] H.Z. Massoud, "Thermal Oxidation of Silicon in Dry Oxygen - Growth Kinetics and Charge Characterization in the Thin Regime," Stanford Electronics Laboratories, TR G502-1, Stanford University, Stanford, Calif., June 1983.

[2.19] R.B. Fair and J.C.C. Tsai, "A Quantitative Model for the Diffusion of Phosphorus in Silicon and the Emitter Dip Effect," *J. Electrochem. Soc.*, **124**, July 1977, pp. 1107-1121.

[2.20] H. Runge, "Distribution of Implanted Ions under Arbitrarily Shaped Mask Regions," *Phys. Stat. Sol. (a)*, **39**, 1977, pp. 595-599.

[2.21] J. Huang and L. Welliver, "on the Redistribution of Boron in the Diffused Layer during Thermal Oxidation," *J. Electrochem. Soc.*, **117**, 1970, pp. 1577-1580.

[2.22] H.G. Lee, R.W. Dutton, and D.A. Antoniadis, "On Redistribution of Boron during Thermal Oxidation of Silicon," *J. Electrochem. Soc.*, **126**, 1979, pp. 2001-2007.

[2.23] M.R. Kump and R.W. Dutton, "An Overview of Process Models and Two-Dimensional Analysis Tools," Stanford Electronics Laboratories, TR G-201-13, Stanford University, Stanford, Calif., July 1982.

[2.24] J.A. Greenfield and R.W. Dutton, "Nonplanar VLSI Device Analysis the Solution of Poisson's Equation," *IEEE Trans. on Electron Devices*, **ED-27**, Aug 1980, pp.1520-1532.

[2.25] R.B. Marcus and T.T. Sheng, "The Thermal Oxidation of Shaped Silicon Surfaces," *J. of Electrochem. Soc.*, **129**, June 1982, pp. 1278-1289.

[2.26] L.O. Wilson, "Numerical Simulation of Gate Oxide Thinning in MOS Devices," *J. Electrochem. Soc.*, **129**, Apr 1982, pp. 831-837.

[2.27] D.Chin, S.Y. Oh, S.M. Hu and J.L. Moll, "Two-Dimensional Modeling of Local Oxidation," presented at Device Research Conference, Colorado, June 1982.

[2.28] E.P. EerNisse, "Viscous Flow of SiO_2," *Appl. Phys. Lett.*, **30**, 1977, pp. 290-293.

[2.29] E.P. EerNisse, "Stress in Thermal SiO_2 during Growth," *Appl. Phys. Lett.*, **35**, 1979, pp. 8-10.

[2.30] D. Chin, S.Y. Oh, R.W. Dutton, and J.L. Moll, "Two-Dimensional Oxidation Modeling," *IEEE Trans. on Electron Devices*, **ED-30**, July 1983, pp. 744-749.

[2.31] D. Chin, S.Y. Oh, R.W. Dutton, and J.L. Moll, "Two-Dimensional Local Oxidation," *IEEE Trans. on Electron Devices*, **ED-30**, Sept 1983, pp. 993-999.

Chapter 3
Device Simulation

3.1 GEMINI : 2-D Poisson Solver

As the dimensions of MOS devices are scaled down, the device structures become more complicated. The insulator/semiconductor interfaces are often non-planar, and the impurity profiles of the devices are complicated and may not be expressed accurately in Gaussian form. The increased complexity of the device structure is necessary for optimization of the device performance, such as minimizing the drain-induced barrier-lowering effects, or enhancing the device reliability, e.g., reducing the electric field at the drain of the MOSFET. Therefore, in the development of VLSI MOS technology, it is essential to be able to simulate the electrical characteristics of devices which have complicated structures. The GEMINI program provides this capability.

The GEMINI program was developed at Stanford University by Greenfield and Dutton [3.1] in 1980. The program performs nonplanar VLSI device analysis by solving the 2-D Poisson equation. The program can accept data from the SUPREM and SUPRA programs, which provide high accuracy in the impurity profile definition, essential for submicron device simulations. The following sections will present in more detail the structure of the program as well as the input format, and then simple examples. In Part B of this book, the use of the GEMINI program is presented in many case studies.

The Capability of GEMINI

Within the 2-D simulation system, the GEMINI program is linked to the SUPREM and SUPRA programs, as well as the PLPKG program for very flexible graphical output. SUPREM provides one dimensional vertical profiles such as the channel and source/drain profiles. This is very useful since the vertical profiles are usually non-Gaussian, an obvious example being source/drain profile. The lateral profile at the source/drain of the MOSFET is specified in the GEMINI input file in this case. If the 2-D process simulator SUPRA is used, then the whole device structure is specified, with the exception of the placement of the electrodes. The SUPRA-GEMINI combination is useful for the simulation of device structures with novel 2-D geometries such as the LDD structure [3.2]. The link between GEMINI and the PLPKG program allows the graphical output of quantities such as potential, impurity, carrier concentrations and electric field distributions, with very flexible format. 1-D, 2-D, as well as bird's-eye-view from different angles are possible. Fig. 3.1 shows schematically the linking of the GEMINI program with the other programs within the 2-D simulation system.

The GEMINI program can simulate device structures such as MOSFET, JFET, MESFET, SOS devices, and other non-planar insulator/semiconductor structures such as the trench isolation structure (this will be described in Chapter 10). Fig. 3.2 shows three examples of structures whose electrical characteristics can be simulated by GEMINI. The structures shown are generated by the GEMINI program. The first one is a "channel length" simulation, where the device structure along the channel is defined. This is the structure that is most often simulated because it calculates the short channel effects where analytical or 1-D approximations are inadequate. The second example is a channel width simulation where narrow width effects such as threshold shifts are simulated. In this case, the capability of simulating non-planar insulator/semiconductor interface is essential. The third example shows an extreme case of a non-planar structure which is the trench isolation in CMOS [3.3] [3.4]. Here, the trench is 5 μm deep and 1 μm wide. All of these

Device Simulation

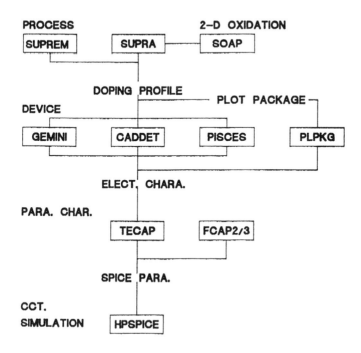

Fig. 3.1. Linkage of GEMINI in 2-D system.

simulations will be presented in detail in Part B of this book.

The program can extract device parameters such as the threshold voltage, subthreshold slope, punchthrough voltage, body effects, and electrode capacitance. Many other applications are also possible and the reader is referred to Part B of this book and also to reference [3.1]. GEMINI has graphic capabilities which allow 2-D plots of the device structure, junctions, depletion edge, as well as 1-D and 2-D contours of quantities such as potential, electric field, carrier concentrations and impurity concentrations. For details, the reader should refer to the GEMINI manual.

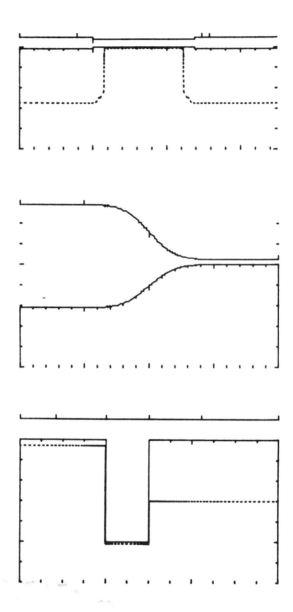

Fig. 3.2. Three major modes of GEMINI inputs.

Device Simulation

Basic Theory

The GEMINI program solves the 2-D Poisson equation only. The Poisson equation is expressed by:

$$\nabla \cdot (\nabla \psi) = -\frac{q}{\varepsilon}(N + p - n) \tag{3.1}$$

where ε is the dielectric constant, ψ is the potential, q is the electron charge, N is the difference between donor and acceptor concentrations, and p and n are the hole and electron concentrations respectively. The electron and hole concentrations are related to the potential and quasi-Fermi potential by the Fermi-Dirac integral of order 1/2. In the program, the potential is defined such that it is equal to the substrate bias in the neutral substrate.

GEMINI does not solve the current continuity equations and it assumes constant quasi-Fermi level within each impurity region. For the MOSFET simulation, the majority carrier quasi-Fermi levels are set equal to the bias in the regions. For the minority carriers, the electron quasi-Fermi potential is set equal to the highest terminal bias and the hole quasi-Fermi potential is set equal to the lowest terminal bias in silicon. Hence the carrier concentration above the threshold voltage is inaccurate for large drain biases. The GEMINI program is useful for the linear and subthreshold regions of the device characteristics. The current calculations in these regions are outlined below.

In the linear region, where the drain to source bias (V_{DS}) is small (<0.1 V), the drain to source current (I_{DS}) is given by the product of the conductance of the channel (G_{DS}) and V_{DS}. G_{DS} is independent of V_{DS} since the current results mainly from minority carrier drift in the presence of the longitudinal electric field induced by the drain bias. The GEMINI program calculates the conductance by assuming zero drain to source bias. The current calculation in this case is valid for arbitrary gate bias. Knowing the device width W and mobility μ, the conductance is calculated by solving for the potential distribution and then determining the channel carrier concentration. For the channel length simulation, the program calculates the quantity Q_L which is defined by

$$Q_L = \frac{1}{\int [\int Q dy]^{-1} dx} \qquad (3.2)$$

where Q is the minority carrier concentration, and x is along the channel and y is vertical to the silicon surface. The y integration is from the silicon surface to the bottom of the substrate. The x integration is between the source and the drain. The current can then be determined by the user, using the expression

$$I_{DS} = q\mu W Q_L V_{DS} \qquad (3.3)$$

where q is the electronic charge. Similarly, for the calculation of the device width, the quantity Q_W is calculated by the program, where Q_W is defined by

$$Q_W = \int [\int Q dy] dx \qquad (3.4)$$

where x is along the channel and perpendicular to current flow and y is vertical to the silicon surface. The y integration is the same as Q_L, and the x integration is between the barriers in both sides. The channel current can be calculated using the expression

$$I_{DS} = \frac{q\mu Q_W V_{DS}}{L} \qquad (3.5)$$

Users should be careful that Q, Q_L, and Q_W in GEMINI are integrated carrier concentrations, not integrated charges.

In the subthreshold region, the current results either from the drain-induced barrier lowering effect, where the applied bias at the drain reduces the barrier height between the source and the drain, or from the lowering of the surface potential by an applied gate bias. In this region, the current can either flow along the interface between the gate dielectric and substrate, or through the substrate. In any case, there exists a point P in the device along the current path where the potential extremum is located. At P the electric field along the direction of current flow vanishes and the current is due to minority carrier diffusion. In this case, the current can be calculated from the potential distribution near this point.

In the calculation of the potential distribution, the minority carrier concentration in the substrate is assumed to be negligible compared with the ionized impurity concentration. Hence in the simulation of the punchthrough

Device Simulation

behavior in the subthreshold region, the results are accurate only for low current levels. However, reasonable accuracy is obtained even when this condition is violated, because the carriers are localized near the current path. The potential distribution, from which the current is calculated, is mainly determined by the device geometry, impurity profile and the applied biases.

GEMINI calculates the quantity Q_B from which the current can be determined. Q_B is given by:

$$Q_B = \frac{Z^*}{L^*} \frac{n_i^2}{N_B} \exp\left[\frac{s(V_s - V_p)}{kT/q}\right] \quad (3.6)$$

where $s = 1$ and -1 for p and n-channel respectively, V_p is the barrier potential, V_s is the source bias voltage, n_i the intrinsic carrier density, N_B the bulk impurity concentration, and Z^* and L^* are the effective width and length of the region of the potential distribution near the point P which controls the current. Fig. 3.3 shows schematically the significance of these two quantities in the case where the current path is through the bulk due to punchthrough. The user can then calculate the drain to source current by using the expression:

$$I_{DS} = -qDWQ_B\left[1 - \exp\left(\frac{sV_{DS}}{kT/q}\right)\right] \quad (3.7)$$

where W is the width of the device, D the minority carrier diffusion constant $(= \mu kT/q)$.

Grid Definition and Numerical Techniques

The grid structure used by GEMINI is rectangular and has nonuniform spacing both in the horizontal and the vertical direction. This grid definition is chosen for accuracy, minimum storage and easy generation. Fig. 3.4 shows a typical grid definition for a channel length simulation. The grid spacings can be defined to be small in regions where the potential changes most rapidly, such as the regions near the source, the drain, and also near the surface. A maximum of 99 grids are allowed for the horizontal as well as the vertical coordinates.

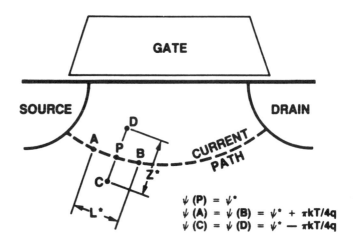

Fig. 3.3. Punchthrough current calculation.

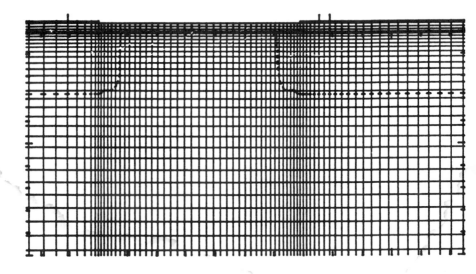

Fig. 3.4. Typical grid for channel-length simulation.

Device Simulation

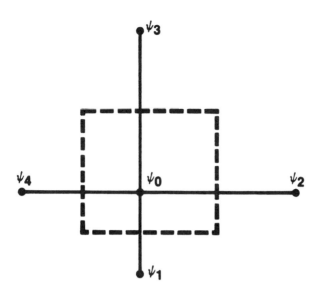

Fig. 3.5. Five-point finite-difference approximation.

Basically, the Poisson equation is solved numerically by using a five point approximation. Fig. 3.5 shows the discretization of the Poisson equation. A rectangular region is defined for each node point by the normal bisectors of the lines joining the node point to its four nearest neighbors. The component of the electric field normal to each side of this rectangle is taken to be constant along that side. The impurity concentration and free carrier concentrations are taken to be constant over the rectangle and are assigned the values obtained by evaluation at the coordinate of that node. For the interface and boundary conditions, the reader should refer to the GEMINI manual.

The discrete form of Poisson's equation consists of a nonlinear system of equations, one for each node point in the rectangular grid. This system of equations is solved by a one-step Newton-SLOR (Successive Line Over-Relaxation) iteration [3.5] [3.6]. To determine if convergence is obtained, the program first locates the point along each vertical grid where the current path is most likely to be located. The potential at this point should be calculated most accurately. This point is defined as the point where the difference between the potential and the majority carrier quasi-Fermi potential is a maximum. The relative changes in the potential at these points are calculated. The iteration is terminated when all of the relative changes are smaller than the

```
COMMENT      NMOS CHANNEL LENGTH SIMULATION
STRUCTURE    TEMPERATURE=300
COMMENT      DEFINE SOLUTION REGION
SUBSTRATE    CONCENTRATION=6E14 P-TYPE WIDTH=5.38 DEPTH=4
COMMENT      DEFINE THE DISCRETIZATION GRID
GRID         XGRID.MIN=0.02 XGRID.MAX=0.05 X.SPACES=90
+            YGRID.MIN=0.01 Y.SPACES=60
COMMENT      DEFINE THE INSULATOR REGIONS
INSULATOR    SOURCE   THICKNESS=0.055 WIDTH=1 ENCROACH=1
INSULATOR    GATE     THICKNESS=0.0375
INSULATOR    DRAIN    THICKNESS=0.055 WIDTH=3 ENCROACH=1
COMMENT      DEFINE THE TOP SURFACE ELECTRODES
ELECTRODE    SOURCE ALUMINUM WIDTH=0.8
ELECTRODE    GATE   N+POLY WIDTH=1.5 LEFTEDGE=1.0
ELECTRODE    DRAIN  ALUMINUM WIDTH=2.8
COMMENT      DEFINE INTERFACE FIXED CHARGE
QSS          CONCENTRATION=2E10
COMMENT      DEFINE THE CHANNEL IMPLANT
PROFILE      CHANNEL IMPLANT SUPRMFILE=DL026
COMMENT      DEFINE THE S-D IMPLANT
PROFILE      SOURCE IMPLANT SUPRMFILE=DSDI5
+            WIDTH=1.0 X.CHAR=0.05
PROFILE      DRAIN  IMPLANT SUPRMFILE=DSDI5
+            WIDTH=3.0 X.CHAR=0.05
END
COMMENT      POISSON SOLUTION DEFINITION
SOLUTION     MAX.ITER=300 DATA.OUT=BD960B
BIAS         SUBSTRATE POTENTIAL=-1
BIAS         SOURCE POTENTIAL=0
BIAS         DRAIN POTENTIAL=3.0
BIAS         GATE POTENTIAL=0.0
END
STEP.VOL     GATE MAXIMUM DELTA.VOL=0.1 STEP=8
+            X.MIN=0 X.MAX=4 Y.MIN=0 Y.MAX=3
+            SUMM.OUT=BS960B
+            MAX.ITER=200
END
```

Fig. 3.6. Input file of GEMINI using SUPREM output.

allowed value.

Examples

In this example, a channel length simulation is presented and the subthreshold current will be calculated. In this case, the combination of SUPREM and GEMINI is used. The coupling between SUPRA and GEMINI will be shown in the next example. The input file for the GEMINI program is discussed briefly. The reader should refer to the manual for details of the input commands.

The input file for the GEMINI program is shown in Fig. 3.6. The input file begins with the structure definition. In this section of the input file, the temperature, device dimensions, substrate, grid, insulators, electrodes, interface charges, and the channel and source/drain profiles are all defined.

In the definition of the substrate, the impurity type and concentration has to be equal to that defined in the SUPREM input file for the simulation of the channel and source/drain profiles. The width and depth defines the region in which the Poisson's equation will be solved. The grid definition allows the user to specify the minimum and maximum grid spacings along the x-direction. The minimum spacing occurs at the left and right edges of the gate insulator region. The maximum spacing occurs at the center of the gate insulator region. The grid spacing changes from minimum to maximum parabolically. The grid spacing outside the gate region is then determined by the total number of horizontal grids specified and the number of grids used up in the gate region. The grid definition for the vertical coordinate consists of a minimum spacing definition, which is at the insulator/semiconductor interface and the total number of grids used for the vertical coordinate. These grid definitions allow the best accuracy in the determination of the potential distribution since more grids are available at the regions where the potential changes most rapidly.

The insulator defines the gate and source/drain region of the device. The thickness and width are defined. Note that GEMINI allows non-planar insulator-semiconductor interfaces. The ENCROACH parameter defines the width of the transition between the different insulator regions when the thicknesses are different. This is essential for the channel width simulation where the field isolation oxide structure has to be defined.

The electrode command defines the materials and positions of the electrodes where biases will be applied to the device. The source and drain electrodes automatically make contact with the diffusion regions. The electrodes cannot overlap in the definition. Fixed charge density at the insulator-semiconductor interface can be specified by the QSS command.

The channel profile can be specified either in the GEMINI input file using Gaussian expressions, or by accepting data from a SUPREM simulation.

The latter method is generally used, since the SUPREM simulation provides the best profile definition. The channel profile will be placed along the gate region, as well as the source and drain region, which is typically what is done in device fabrication. The source and drain profiles are then specified by first defining the width of these regions. Then the data file name of the SUPREM simulations are specified. Since SUPREM is a one dimensional profile, the lateral diffusion of the source and drain profiles have to be defined in GEMINI. In this case, the characteristic length X.CHAR of a lateral Gaussian distribution is defined. The structure definition for the device is ended by an END command.

The next section of the input file is the solution definition, where the bias conditions are specified. In this example, the substrate, source, drain, and gate biases are specified. If the device definition has an error which causes non-convergence, the MAX.ITER specifies the maximum number of iterations allowed. For a MOSFET simulation, the number of maximum iterations needed is typically 100 or less. The output data file name which contains the solution of the Poisson's equation are also defined. The END command ends the solution definition.

The next section of the input file is for parameter extraction. The subthreshold characteristics of the device can be generated by stepping the gate bias and calculating the quantity Q_B, from which the current can be determined. The STEP.VOL definition specifies the voltage to be stepped, which is GATE in this example. The voltage increment DELTA.VOL and the number of steps STEP are specified. The MAXIMUM parameter defines the path where the current will flow. For this example, the carriers are electrons. Hence the current path is located where the potential is a MAXIMUM. For a p-channel MOSFET simulation, the parameter would be MINIMUM. The approximate region where the current path will be located is specified to minimize the search time for the current path. The summary output file name is specified here. The summary consists of the applied voltages as well as the quantity Q_B as a function of the gate voltage. The END command ends the STEP.VOL definition.

With the SUPREM data files available and the GEMINI input file completed, the program can be executed and the solution plotted in various

Device Simulation

formats.

Fig. 3.7 shows the 2-D plot of the device structure and also the potential profiles with the device biased as defined in the SOLUTION definition. By using the PLPKG program, the bird's eye view of the impurity and potential profiles are plotted and shown in Fig. 3.8 and 3.9. The calculated device subthreshold characteristics are plotted in Fig. 3.10. From this graph, the subthreshold slope and threshold voltage (as defined by a threshold current) can be determined.

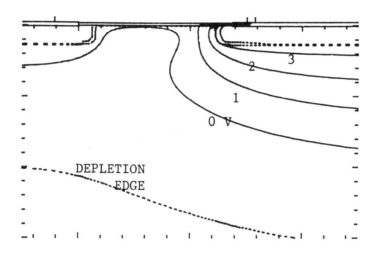

Fig. 3.7. Device structure and 2-D potential distribution of the example.

The second example briefly shows the results of the SUPRA and GEMINI combination in the channel width simulation. The reader is referred to Chapter 11 for more details of using SUPRA-GEMINI for calculating narrow width effects. Fig. 3.11 shows the input file for the GEMINI program. In this case, the input file is very simple since the structure is essentially defined by the SUPRA data file. The only structure input in this example is the gate definition. The SOLUTION and STEP.VOL definitions are the same as the SUPREM-GEMINI combinations. Fig. 3.12 shows the structure of the device, including the 2-D impurity profile, and also the depletion edge calculated by

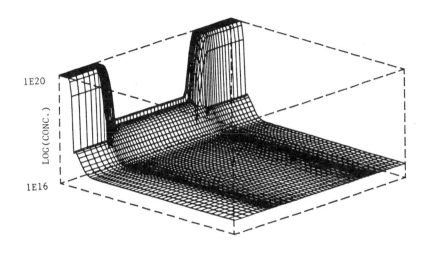

Fig. 3.8. Bird's-eye-view of impurity distribution.

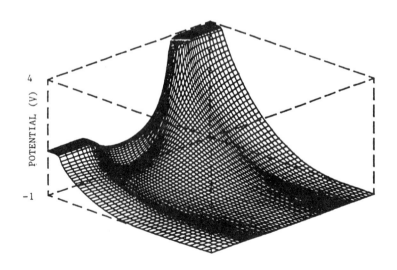

Fig. 3.9. Bird's-eye-view of potential distribution.

Device Simulation

the GEMINI program. One disadvantage in the SUPRA-GEMINI combination is that the grid is defined as in the SUPRA simulation, which is often not optimized for potential calculations. Fig. 3.13 shows the I_{DS}-V_{GS} characteristics of this device simulated by the GEMINI program. This type of channel width simulation is very useful for the development of isolation technologies where the narrow width effects must be minimized.

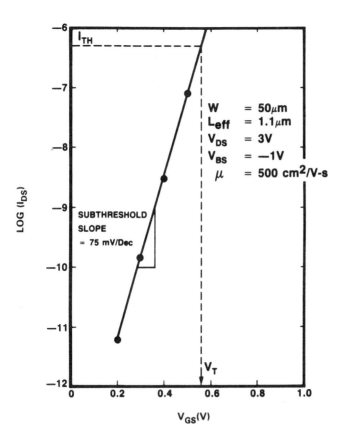

Fig. 3.10. N-channel MOSFET Subthreshold characteristics simulated by GEMINI.

```
COMMENT LOCOS ISOLATION CHANNEL WIDTH SIMULATION
STRUCTURE DATA.INP=GSW76
ELECTRODE GATE N+POLY   WIDTH=0.99   LEFTEDGE=1
END
SOLUTION  MAX.ITER=300   DATA.OUT=LOS78
BIAS      SUBSTRATE   POTENTIAL=0
BIAS      GATE        POTENTIAL=1
END
```

Fig. 3.11. GEMINI Input file for the channel width simulation.

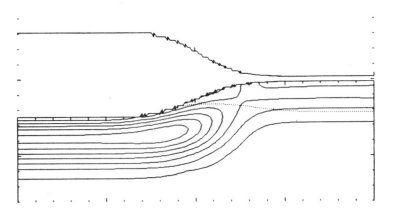

Fig. 3.12. Device structure of channel width simulation.

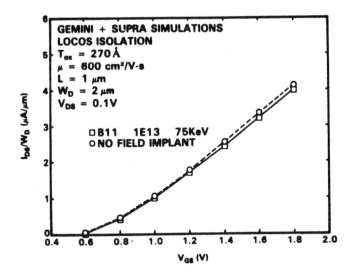

Fig. 3.13. Linear region characteristics for the narrow width devices.

3.2 CADDET : 2-D 1-Carrier Device Simulator

As shown in the previous section, GEMINI can simulate a MOSFET when either V_{DS} is small or either the channel carrier density is so small that the potential distribution is not disturbed as is the case with the subthreshold or weak punchthrough regions. It cannot be used, however, when the channel carrier density is larger than the fixed charge density or when V_{DS} is not small as is the case in most of the triode region and all the saturation region. In order to simulate the whole range of the semiconductor device, Poisson, electron and hole continuity equations should be solved with proper boundary conditions in two dimensions. In a field-effect device, most of the current is carried by the majority carrier of the source/drain. Thus, field-effect devices can be simulated by solving only the majority carrier current continuity equation and Poisson's equation to save calculation time. CADDET is a dedicated program for the field-effect transistor and functions in this way. It can simulate the whole range of the field-effect transistor (subthreshold, linear, saturation and punchthrough) with reasonable CPU time.

CADDET can analyze planar JFET's and MOSFET's with various channel, source, and drain configurations. Fig. 3.14 illustrates the input structures of the field-effect transistors that can be simulated by CADDET. The only limitation is that the silicon and oxide surface must be planar. The device structure is selected by the structure name and the impurity profiles are specified by the step and Gaussian profiles. For these field-effect device structures, the whole range of operation can be simulated and the terminal currents can be calculated without any limitation on bias. CADDET has been linked to the PLPKG program so that the distributions of the electrons, holes, impurities, and potential can be displayed in 1-D, 2-D and 3-D plots. The distribution of the field and velocity can also be plotted. CADDET is widely used to analyze effects of velocity saturation, break-down, hot electron generation, punchthrough, and the Lightly Doped Drain (LDD) structure. An analysis of the LDD structure will be presented in detail in Part B. CADDET is also indispensable for generating the whole I-V characteristics in order to extract SPICE parameters for the circuit simulations.

Device Simulation

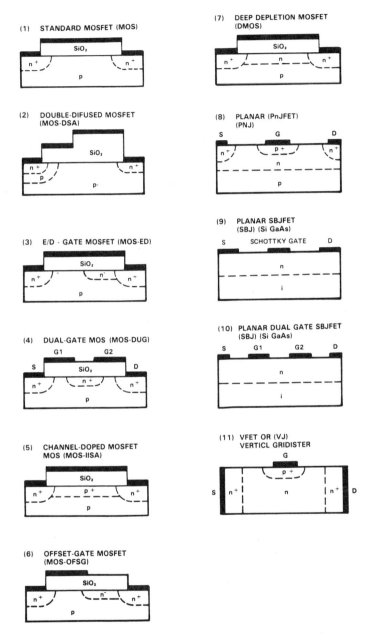

Fig. 3.14. Input structures of CADDET.

There are several limitations in CADDET due to the simplifications. First, only n-channel devices can be simulated because only the electron continuity equation is solved. For p-channel devices, the structure should be converted to the equivalent n-channel device and then simulated. The simulation results are then converted to the p-channel characteristics. Device structures with planar surfaces and Gaussian profiles are approximations of the real device structures. Though CADDET is adequate for many applications, the simplifications of the structure and impurity profile can be a source of inaccuracy in scaled submicron devices, where the effects of the non-planar structure and non-Gaussian profiles are significant. For more details on CADDET, refer to the CADDET manual [3.7].

Basic Equation and Numerical Algorithm

To cover the whole range of the field-effect transistor operation, CADDET solves the Poisson and electron continuity equations. The Poisson equation is

$$\nabla^2 \psi = -\frac{q}{\varepsilon}(p - n + N_D^+ - N_A^-) \tag{3.8}$$

where ψ is the electrostatic potential and ε is the dielectric constant; N_D^+ and N_A^- are ionized donor and acceptor concentrations respectively; n and p are electron and hole concentrations respectively. Although Fermi-Dirac statistics should be used in the semiconductor, Maxwell-Boltzman statistics is employed instead, because it is simple and is still a good approximation. Then the electron and hole densities can be expressed by

$$n = n_i \exp\left(\frac{\psi - \phi_n}{kT/q}\right) \tag{3.9a}$$

$$p = n_i \exp\left(\frac{\phi_p - \psi}{kT/q}\right) \tag{3.9b}$$

where n_i is the intrinsic carrier density; T is the absolute temperature, and k is the Boltzman constant; ϕ_n and ϕ_p are the quasi-Fermi potentials of electrons and holes respectively.

Device Simulation

The electron continuity equation is formulated using the stream function [3.8] as a basic variable instead of electron and hole concentrations to enhance numerical stability. When the recombination and generation of carriers are neglected, which is reasonable in MOSFETs, the divergence of the electron current density is zero. The electron current density, therefore, can be a curl of some vector (stream function Θ). The current density ($\mathbf{J}$) is formulated as below when the normalization [3.9] and Einstein's relation are used.

$$\nabla \cdot \mathbf{J}_n = 0 \qquad (3.10a)$$

$$\mathbf{J}_n = \mu_n e^\psi \nabla(n e^{-\psi}) \qquad (3.10b)$$

$$\mathbf{J}_n = J_o \nabla \times \Theta \qquad (3.10c)$$

For a two-dimensional case, the stream function has only the z component, θ. The x and y components of electron current density can be written

$$J_x = \mu_n e^\psi \frac{\partial}{\partial x}(n e^{-\psi}) = J_o \frac{\partial \theta}{\partial y} \qquad (3.11a)$$

$$J_y = \mu_n e^\psi \frac{\partial}{\partial y}(n e^{-\psi}) = -J_o \frac{\partial \theta}{\partial x} \qquad (3.11b)$$

Divide both sides of the above equations by $\mu_n e^\psi$, then differentiate the first equation partially with respect to y, the second equation partially with respect to x, and add the two obtaining

$$\frac{\partial}{\partial x}\left[\mu_n^{-1} e^{-\psi} \frac{\partial \theta}{\partial x}\right] + \frac{\partial}{\partial y}\left[\mu_n^{-1} e^{-\psi} \frac{\partial \theta}{\partial y}\right] = 0 \qquad (3.12)$$

This equation is equivalent to the current continuity Eq. (3.10a). After θ is solved with appropriate boundary conditions, J_o should be evaluated using the following equation based on Eq. (3.11a).

$$J_o = \frac{-N_D^+(1 - e^{-V_{DS}})}{\int_0^{L_{eff}} \mu_n^{-1} e^{-\psi} \frac{\partial \theta}{\partial y} dx} \qquad (3.13)$$

The integration in the denominator is along the channel. The electron density is calculated using Eqs. (3.11a) and (3.11b) with J_o and θ.

Eqs. (3.8) and (3.12) are discretized by the standard 5-point finite-difference approximation. A non-uniform grid is employed to enhance accuracy. The mesh is automatically generated by the program as shown in Fig. 3.15. In the vertical (y) direction, a fine grid is used at the silicon surface and grid spacing increases geometrically toward the substrate. Fine grids are used in the channel and source/drain junctions and coarse grids are used in the middle of the channel in the horizontal (x) direction. The user can specify the total number of grids, the minimum spacing and geometrical ratio for both x and y directions. The maximum number of grid points is limited to 2000. The meshes for potential and for stream function are different and are interleaved as shown in Fig. 3.16. The circle is the node for the potential and the cross is the node for the stream function. The stream function mesh is located in the middle of the potential mesh. To reduce the calculation time, these equations are solved iteratively (Gummel iteration) after discretization instead of by the simultaneous solution of both equations. The flow chart of the program is illustrated in Fig. 3.17. When the program starts, the grid is generated, the potential, hole and electron quasi-Fermi potentials are initialized, and the equations are discretized. The hole quasi-Fermi potential is set constant throughout the device and is the same as that of the substrate. First, the

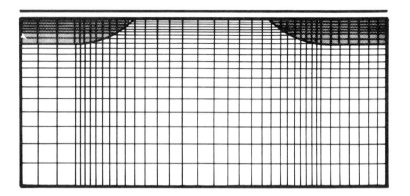

Fig. 3.15. Non-uniform rectangular grid in CADDET.

Device Simulation

Poisson equation is discretized and solved by Stone's method [3.10] for the potential with the fixed electron quasi-Fermi potential. Then, the electron continuity equation is discretized and solved by SLOR (Successive Line Over Relaxation) for the electron stream function with fixed potential. The electron densities and electron quasi-Fermi potential are calculated from the stream function and are updated. This procedure is repeated until the solutions converge. Finally, the drain current is calculated from the converged stream function.

Mobility Model

In MOS devices, the drain current is directly proportional to the mobility. Because an accurate mobility model is very important for the drain current calculation in the linear and saturation regions, it is presented in detail in this section.

Mobility depends on both the parallel field and the doping density in the bulk silicon. In MOS devices, it also depends on the normal field. The original

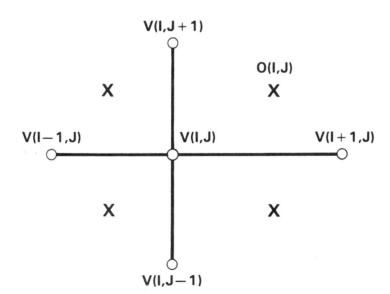

Fig. 3.16. Meshes of potential and stream functions.

mobility model in CADDET is based on Gummel's bulk mobility model [3.11] for the parallel field and doping dependence, and an empirical model for the normal field dependence. The model equations are shown below [3.12]

$$\mu_n = \mu_o f(N_B, E_p) g(E_n) \tag{3.14a}$$

$$f(N_B, E_p) = \left[1 + \frac{N_B}{N_B/S + N} + \frac{(E_p/B)^2}{N_B/S + N} + (E_p/B)^2\right]^{-1/2} \tag{3.14b}$$

$$g(E_n) = (1 + \alpha E_n)^{1/2} \tag{3.14c}$$

where N_B is the impurity density; E_p and E_n are the field components parallel and normal to the current-flow, respectively; μ_o is 1400 cm^2/V sec, and S is

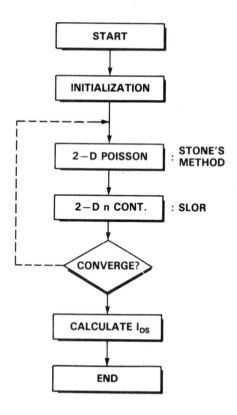

Fig. 3.17. Flow chart of CADDET.

Device Simulation

350; N is 3×10^{16} cm^{-3} and B is 7.4×10^3 V/cm; α is 1.54×10^{-5} cm/V. This model, however, doesn't agree well with measurements, especially for thin gate oxide and the highly doped channel. In Fig. 3.18, simulations with this model are compared with measurements. The thickness of the gate oxide is

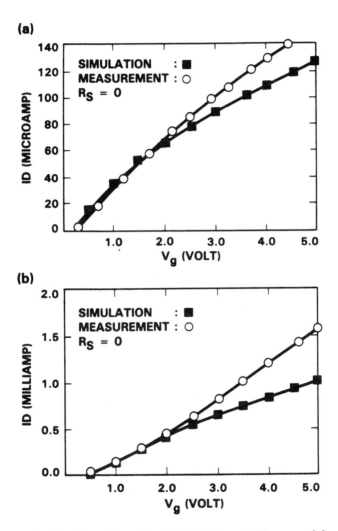

Fig. 3.18. Bench-mark of CADDET mobility model. (a) $V_{DS} = 0.1$ V (b) $V_{DS} = 5$ V.

25 nm and the effective channel length is 0.9 μm. The drain voltage is 0.1 V in Fig. 3.18(a) and 5 V in Fig. 3.18(b). Disagreement is severe in the saturation region, due to the way the normal field degradation is combined with the bulk mobility model. The normal field not only degrades the zero parallel field mobility but also degrades the saturation velocity. This does not agree with experimental observations.

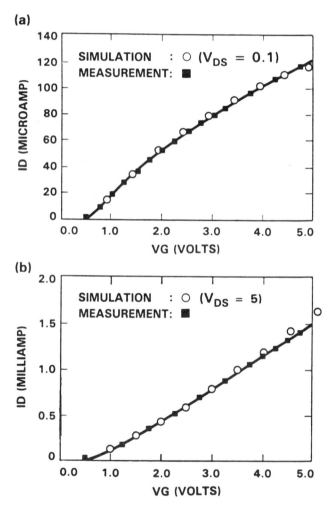

Fig. 3.19. Bench-mark of the new mobility model.

Device Simulation

A new mobility model has been developed. It is based on a modified Gummel's model [3.13] and on a newly characterized normal field mobility [3.14]. The normal and parallel field dependencies are combined in such a way to keep the saturation velocity constant to obey Thornber's scaling rule [3.15]. The new model is

$$\mu_o(E_n, E_p) = \frac{\mu_o(E_n)}{\left\{1 + \frac{[\mu_n(E_n)E_p/V_c]^2}{\mu_o(E_n)E_p/V_c + G} + \left[\frac{\mu_o(E_n)E_p}{V_s}\right]^2\right\}^{1/2}} \quad (3.15a)$$

$$\mu_o(E_n) = 690 \, (E_n/10^5)^{-0.28} \text{ cm}^2/\text{V sec} \quad (3.15b)$$

where V_s is 1.036×10^7 cm/sec, V_c is 4.9×10^6 cm/sec, G is 8.8. The normal field, E_n, is in volt/cm. The comparisons of the new model and the measurement are illustrated in Fig. 3.19. Agreement is good in the linear and the saturation regions. CADDET simulations with the new model have been bench-marked for a wide range of device parameters. Agreements are within 10%. CADDET with the new mobility model enables accurate simulation of a wide range of MOS devices with reasonably short computation time. In the next section, the simulation of a conventional MOS device will be illustrated as an example of CADDET applications.

N-channel MOSFET simulation by CADDET

In CADDET, several MOSFET and JFET structures can be simulated such as conventional MOS, DMOS, p-n junction FET, Schottkey-barrier JFET and vertical FET as shown in Fig. 3.14. In this section, the simulation of the conventional MOSFET will be illustrated as a simple case study to show how to use CADDET. Fig. 3.20 shows the structure of the n-channel MOSFET to be simulated. The gate oxide thickness is 46.8 nm and the effective channel length is $1.0 \, \mu\text{m}$. The junction depth of the source/drain is $0.3 \, \mu\text{m}$ and the substrate doping is 6×10^{14} cm^{-3}.

Before running the program, an input file must be prepared which describes the device structure and doping distribution. The input for the above device is shown in Fig. 3.21. A fixed format input is employed. The problem with this kind of input is that it is easy to make errors and it is difficult to

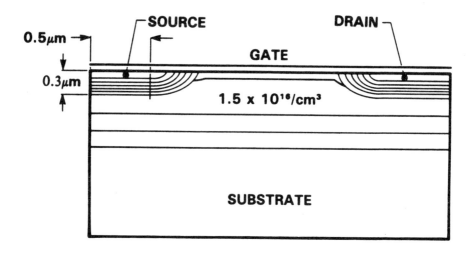

Fig. 3.20. Device structure of n-channel MOSFET for CADDET simulations.

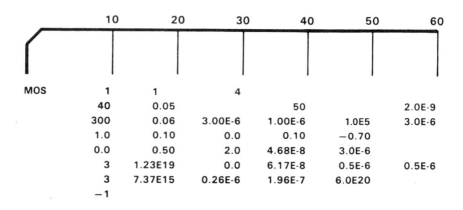

Fig. 3.21. CADDET input for the example device.

Device Simulation

read and understand. The input consists of 80 column lines. A line is divided into fields. The first line is divided into 5-column fields and the rest is divided into 10-column fields. One feature of CADDET that differs from other device simulators is in the unit system. CADDET uses the MKS system. Device type and other control parameters are specified in the first line. "MOS" in the first field designates the device type as a conventional MOSFET. The second field specifies the number of gates as one. The second line specifies the numbers of x and y grids and their spacings. The third line is the device parameters. Temperature is in the first field as $300\,°K$, mobility in the second field as $0.06\,m^2$/volt sec, device length in the third field as $3\,\mu m$, channel width in the fourth field as $1\,\mu m$, and the substrate depth in the sixth field as $3\,\mu m$. In the fourth line, the drain biasing schedule is described. The first parameter gives the back gate bias as -1 volt. In CADDET, the back gate bias is specified as V_{SB}, that is, the source bias with respect to the substrate. Be careful! It is the negative of the conventional back gate bias. The second parameter is the starting drain bias. The stepping voltage and the final bias of V_{DS} are specified in the third and fourth fields, respectively. The drain bias is fixed at 0.1 volt in this example. The flatband voltage is in the fifth field. The fifth line is the gate biasing. The initial gate bias is in the first field, the voltage step in the second field, and the final gate bias in the third field. In this example, V_{GS} starts from 0 V and stepped by 0.5 V to 2.0 V. The fourth and fifth parameters specify the gate oxide thickness and gate length, respectively.

The sixth line describes the doping distribution of the source/drain. In CADDET, the source/drain profile can be step, Gaussian or implanted Gaussian. The first parameter in the fifth line gives the doping type as implanted Gaussian. The profile shape and equation in the implanted Gaussian are illustrated in Fig. 3.22. The dose, range, and standard deviation are specified in the second, third and fourth fields respectively. The fifth and sixth parameters specify L_S and L_D respectively. The channel profile is in the seventh line. The first parameter is the type of the channel profile. It can be a uniform, a redistributed or an implanted profile. The implanted profile is selected in this case and illustrated in Fig. 3.23. The dose $(N_p\,')$, range $(R_p\,')$ and standard deviation$(\sigma\,')$ are in the second, third and fourth fields, respectively. The fifth parameter is the substrate doping (N_A). In the last line, "-1" signals the end of the input. After preparing the input, CADDET can be run either in the interactive or batch mode. When CADDET is run, it first asks for the input

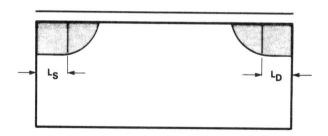

$$N_1(i,j) = \begin{cases} \dfrac{N_p}{\sqrt{2\pi}\sigma} \exp\left[-\dfrac{(y_j - R_p)^2}{2\sigma^2}\right] & \left(0 \le x_i \le L_S \text{ or } L_x - L_D \le x_i \le L_x\right) \\ \dfrac{N_p}{\sqrt{2\pi}\sigma} \exp\left[-\dfrac{(x_i - L_S)^2 + (y_j - R_p)^2}{2\sigma^2}\right] + \dfrac{N_p}{\sqrt{2\pi}\sigma} \exp\left[-\dfrac{(x_i(L_x - L_D))^2 + (y_j - R_p)^2}{2\sigma^2}\right] \\ & (L_S \le x_i \le L_x - L_D) \end{cases}$$

Fig. 3.22. Implanted Gaussian profile for the source/drains.

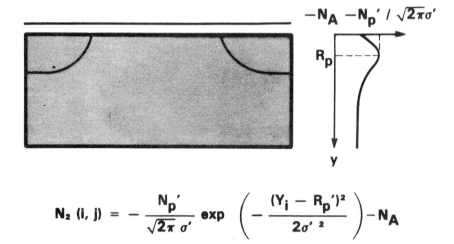

$$N_2(i,j) = -\dfrac{N_p'}{\sqrt{2\pi}\sigma'} \exp\left(-\dfrac{(Y_i - R_p')^2}{2\sigma'^2}\right) - N_A$$

Fig. 3.23. Implanted Gaussian profile for the channel.

Device Simulation

0*	VGS	VDS	VBG	ID
	0.000	.100	0.000	3.666E-14
	0.500	.100	0.000	2.572E-09
	1.000	.100	0.000	8.693E-07
	1.500	.100	0.000	2.281E-06
	2.000	.100	0.000	3.592E-06
	2.500	.100	0.000	4.810E-06
	3.000	.100	0.000	5.950E-06
	3.500	.100	0.000	7.039E-06
	4.000	.100	0.000	8.079E-06
	4.500	.100	0.000	9.080E-06
	5.000	.100	0.000	1.005E-05

Fig. 3.24. Summary file of CADDET.

file name. Then it reads the input file and echoes the input to the terminal. After that, it starts to iterate to solve the Poisson and electron continuity equation. It prints out the results of each iteration. The drain currents and bias conditions are summarized and saved in the summary file as shown in Fig. 3.24.

3.3 PISCES-II : General-Shape 2-D 2-Carrier Device Simulator

PISCES-II is a full 2-D semiconductor device simulation program which has been developed by Stanford University. It solves the Poisson equation and current continuity equations for up to two carriers in two dimensions to simulate the electrical characteristics of devices under either steady-state or transient conditions. The program solves these equations on non-uniform triangular grids so that the device structure can be completely arbitrary with general doping profiles, obtained either from analytical functions or SUPREM-III. The details of the physical models and input syntax are described in PISCES-II user's manual [3.17].

PISCES-II has many applications. It is ideally suited to simulate and study new device structures because it solves all governing equations in the semiconductor with very few approximations. Combined with SUPREM-III, it can be used in the early phases of process and device development to design process experiments and understand device operation and problems. It will reduce the development cost and time. PISCES is also useful to examine the sensitivity of device electrical parameters with respect to the device geometry and process parameters.

In contrast to the limited device operation region and geometry of GEMINI, PISCES-II provides simulations of full-range operation of any homogeneous semiconductor device with arbitrary geometry and doping profile. There are several problems, however. An arbitrary geometry inevitably complicates the grid generation. The inclusion of the current continuity equation brings the possibility of non-convergence. Thus, a user needs some knowledge of optimal grid generation and solution methods to avoid non-convergence. In the worst case, the user needs to adjust the grid and experiment with solution methods to solve the convergence problem.

GEMINI has a simple input for the device structure and grid. It seldom has convergence problems because it solves only the Poisson equation. Considering the fact that the device parameters for the typical MOSFET needed by process engineers (threshold voltage(V_T), subthreshold slope, leakage current and punchthrough voltage) can be simulated by GEMINI, it is understandable that many process engineers still prefer GEMINI. However, the progress of

Device Simulation

device technology to sub-micron dimensions has extended the required knowledge of devices in a new technology beyond the capability of GEMINI. There is a great need to migrate to a more general analysis program such as PISCES-II. A difficulty, however, is that users consider it as a time-consuming job. Moreover, PISCES-II needs several key enhancements which will make it more useful. A mobility model that is better than the default model should be implemented to provide the accuracy acceptable for generation of I-V characteristics and SPICE parameter extractions. For the study of hot electron related reliability problems, an impact-ionization model is needed.

To practically eliminate the convergence problem and make it easy for users to begin with or migrate from GEMINI to PISCES-II, a terminated rectangular grid by eliminations has been developed and tested. A parallel- and vertical-field dependent mobility model is implemented. It enables the accurate simulation of the linear and saturation operation of MOSFETs. With these provisions, PISCES-II will be as simple as GEMINI but with flexible geometry and more capabilities. PISCES will calculate the drain current and V_T correctly and eliminate the complication of the quasi-fermi level specification in GEMINI because the quasi-fermi level is calculated from the current continuity equations.

A simple tutorial will be presented on the application of PISCES-II to a n-channel device. All the doping profiles used are generated by SUPREM-III.

Basic equations and numerical algorithm

The operation of a semiconductor device can be completely specified by both Poisson equation and the electron and hole continuity equations with proper boundary conditions. The basic equations are

$$\nabla \cdot (\nabla \psi) = -\frac{q}{\varepsilon}(N + p - n) \qquad (3.16a)$$

$$\nabla \cdot \mathbf{J}_n = qU(n,p) + q\frac{\partial n}{\partial t} \qquad (3.16b)$$

$$\nabla \cdot \mathbf{J}_p = -qU(n,p) - q\frac{\partial p}{\partial t} \qquad (3.16c)$$

where

$$\mathbf{J}_n = q\mu_n \left(-n\nabla\psi + \frac{kT}{q}\nabla n \right) \tag{3.16d}$$

$$\mathbf{J}_p = q\mu_p \left(-p\nabla\psi - \frac{kT}{q}\nabla p \right) \tag{3.16e}$$

The Poisson equation governs the electrostatic potential and the electron and hole continuity equations govern the carrier concentrations. N is the net concentration of ionized impurity. $U(n,p)$ represents net electron and hole recombination. PISCES-II supports both Shockley-Read-Hall and Auger recombination. μ_n and μ_p are electron and hole mobilities. The same electron mobility model used in CADDET has been implemented in PISCES-II. A similar model for hole mobility has been developed and implemented. These mobility models are

$$\mu(E_n, E_p) = \frac{\mu_o(E_n)}{\left\{ 1 + \frac{[\mu_o(E_n)E_p/V_c]^2}{\mu_o(E_n)E_p/V_c + G} + \left[\frac{\mu_o(E_n)E_p}{V_s} \right]^2 \right\}^{1/2}} \tag{3.17}$$

where E_p is the field parallel to the current flow and E_n is the field normal to the current flow.

For electrons, the E_n dependence and coefficients are

$$\mu_o(E_n) = 690 \left(\frac{E_n}{10^5} \right)^{-0.28} \text{ cm}^2/\text{V sec} \tag{3.18a}$$

$$V_s = 1.036 \times 10^7 \text{ cm/sec} \tag{3.18b}$$

$$V_c = 4.9 \times 10^6 \text{ cm/sec} \tag{3.18c}$$

$$G = 8.8 \tag{3.18d}$$

For holes, the E_n dependence and coefficients are

$$\mu_o(E_n) = (828.56 - 55.77 \log E_n) \text{ cm}^2/\text{V sec} \tag{3.19a}$$

$$V_s = 1.2 \times 10^7 \text{ cm/sec} \tag{3.19b}$$

$$V_c = 2.128 \times 10^6 \text{ cm/sec} \tag{3.19c}$$

$$G = 1.6 \tag{3.19d}$$

To solve the governing equations on a computer, they must be discretized on a grid. In PISCES-II, each equation is integrated over a small polygon enclosing each node, which is formed by the bisectors of the lines connecting to the neighboring nodes. Using current and electric flux conservation, the integration equates the net flux into the polygon with the sources and sinks inside it. If the number of nodes is N, this method generates $3N$ non-linear algebraic equations for unknown potentials and carrier concentrations. To avoid negative concentrations, the Gummel-Scharfetter formula [3.11] is used to discretize the current flux in the continuity equations. For further details of discretization, see Price's thesis [3.17].

By discretization, the semiconductor device equations are transformed into $3N$ coupled non-linear algebraic equations. These equations should be solved by a non-linear iteration method. There are two approaches; one is the decoupled method (Gummel method) and the other is the coupled method (Newton method). In the Gummel method, these equations are solved sequentially. First, the Poisson equation is solved for potential with fixed quasi-fermi levels. Then, the new potential is substituted into the continuity equations. The new carrier concentrations are fed-back into the Poisson equation. The same procedure is repeated until it converges. In this way, only one equation is solved at a time. Thus, the matrix size is only N. This decoupled method converges quickly when the interaction of the Poisson and continuity equations is small such as in the subthreshold region or when V_{DS} is small. However, the convergence slows down when the interaction is significant such as in the linear and saturation regions of a MOSFET. ICCG (Incomplete Cholesky Conjugate Gradients) iterative matrix solution method is recommended to solve the equations in the Gummel method. It is the fastest iterative method known, which can be used on an irregular grid.

In the Newton method, all the equations are solved simultaneously. The single matrix equation includes all the coupling between the equations. Due to this, the Newton algorithm converges at a rate that is nearly independent of bias conditions. The size of the matrix, however, is 3 times as large (or twice as large when solving for only one carrier) as the matrix size of the Gummel method. Thus, the cpu time per iteration is long but the number of iterations is smaller. The single biggest acceleration of a Newton iteration is the

Newton-Richardson procedure which only refactors the Jacobian matrix when necessary. With this procedure, the Newton method with one-carrier is faster than the Gummel method in the linear and saturation regions of a MOSFET. The direct matrix solution method is recommended for the Newton method.

At the beginning of the iterative process, PISCES must have an initial guess for the potential and carrier concentrations at each node point. The initial guess directly affects the rate of convergence. There are four types of initial guesses used in PISCES-II. The first is **init** initial guess which uses the charge neutral assumption and is only valid when V_{DS} is zero. Any later solution begins with an initial guess obtained by modifying one or two previous solutions. For a **previous** initial guess, the previous solution is used as an initial value only modified by changing the applied bias at the contacts. Sometimes, a better guess can be obtained by **local** guess. In this case, the previous solution is used and the majority fermi levels is changed to be equal to the bias applied at the contact connected to this region. Frequently, the best initial guess is obtained by **projection**. In this method, two previous solutions with appropriate bias conditions are used and extrapolated to the new bias condition. In a MOSFET, the **projection** method yields rapid convergence when the gate bias is stepped except near the subthreshold and linear region transition. The **previous** works better than **local** or **projection** when the drain bias is stepped.

Example

An n-channel transistor structure is used for this example. The grid generation has been implemented through macros so that they are transparent to the users. The information that must be supplied by the user is similar to that required by GEMINI. There are three phases of PISCES-II simulations. First is the specification of the device structure. Doping profiles should be specified and optimum grid should be generated. All information is saved in a mesh file. Then, device characteristics at specified bias conditions are simulated and the solutions are saved. Finally, graphical displays of I-V characteristics or internal distributions are performed to extract the desired information from the solutions. The detailed input of grid generation (expanded by a macro processor) will be explained.

Grid generation

The correct generation of grid is a critical issue in device simulations. To minimize the discretization errors, the grid should be fine enough to resolve rapid changes of internal distributions. The grid should also be a reasonable fit to the device shape to accurately represent the device geometries. However, the number of nodes in a grid should be minimized because simulation time is proportional to N^α where α usually varies between 1.5 and 2. Above all, inadequate grid is the main cause of the non-convergence in PISCES-II. To satisfy the above conditions, PISCES-II adopts a flexible triangular grid structure. The most important disadvantage of this grid is the difficulty of user specification. To overcome this, PISCES-II supports two grid generation methods and automatic regrid capability. The first is the rectangular grid generated using input statements. The second is a stand-alone interactive grid generator. Based on our experience with the automatic regrid, it is difficult to control the number and location of nodes. It also generates obtuse triangles in critical areas, such as the channel in a MOSFET structure. These obtuse triangles can disturb the solution accuracy and in the worst case cause convergence problems. Thus, the terminated rectangular grid with elimination is recommended for planar MOS structures. The PISCES-II input of this grid will be illustrated for a n-channel device. The full input from the macro processor is shown in Fig. 3.25. The following will explain the input file in detail.

```
title cmos n-channel device : n4a.i
$
$*** define NMOSGRD for device structure and regrid
option  hp2623 x.s=7
$
$ mosgrid x.dim=3 y.dim=3 tox=0.025 x.1=0.9 x.4=2.1 meshf=n4msh3
$
$ *** DEFINE THE RECTANGULAR MESH ***
MESH   RECT  NX=37  NY=34  OUTF=n4msh3
```

```
1... title CMOS n-channel device : n4a.i
2... $
3... $***  define NMOSGRID for device structure and regrid
4... option    x.s=7
5... $
6... $ mosgrid x.dim=3 y.dim=3 tox=0.025 x.1=0.9 x.4=2.1 meshf=n4msh3
7... $
8... $ **** DEFINE THE RECTANGULAR MESH ****
9... $
10... MESH      RECT      NX=37 NY=33 OUTF=n4msh3
11... $
12... X.M       N= 1      L=0.0000      R=1.0000
13... X.M       N= 6      L=0.8100      R=0.7000
14... X.M       N= 7      L=0.8500      R=1.0000
15... X.M       N=12      L=0.9500      R=1.0000
16... X.M       N=13      L=0.9800      R=1.0000
17... X.M       N=19      L=1.5000      R=1.1000
18... X.M       N=25      L=2.0200      R=0.9091
19... X.M       N=26      L=2.0500      R=1.0000
20... X.M       N=31      L=2.1500      R=1.0000
21... X.M       N=32      L=2.1900      R=1.0000
22... X.M       N=37      L=3.0000      R=1.6000
23... $
24... Y.M       N= 1      L=-0.0250     R=1.0000
25... Y.M       N= 4      L=0.0000      R=1.0000
26... Y.M       N= 5      L=0.0005      R=1.0000
27... Y.M       N= 6      L=0.0010      R=1.0000
28... Y.M       N= 7      L=0.0020      R=1.0000
29... Y.M       N= 8      L=0.0040      R=1.0000
30... Y.M       N= 9      L=0.0080      R=1.0000
31... Y.M       N=10      L=0.0160      R=1.0000
32... Y.M       N=11      L=0.0320      R=1.0000
33... Y.M       N=19      L=0.3000      R=1.1000
34... Y.M       N=23      L=0.5000      R=1.1000
35... Y.M       N=27      L=1.0000      R=1.1000
36... Y.M       N=33      L=3.0000      R=1.2000
37... $
38... ELIM      Y.DIR     IX.L=26   IX.H=32   IY.L=19   IY.H=33
39... ELIM      Y.DIR     IX.L=7    IX.H=12   IY.L=19   IY.H=33
40... ELIM      Y.DIR     IX.L=13   IX.H=16   IY.L=21   IY.H=33
41... ELIM      Y.DIR     IX.L=22   IX.H=25   IY.L=21   IY.H=33
42... ELIM      Y.DIR     IX.L=6    IX.H=32   IY.L=24   IY.H=33
43... ELIM      Y.DIR     IX.L=1    IX.H=34   IY.L=28   IY.H=33
44... $
45... ELIM      X.DIR     IX.L=1    IX.H=5    IY.L=5    IY.H=11
46... ELIM      X.DIR     IX.L=33   IX.H=37   IY.L=5    IY.H=11
47... $
48... REGION    NUM=1     IX.L=1    IX.H=37   IY.L=1    IY.H=4    OXIDE
49... REGION    NUM=2     IX.L=1    IX.H=37   IY.L=4    IY.H=33   SILICON
50... $
51... $ **** ELECTRODES : 1-DRAIN  2-GATE  3-SOURCE  4-SUBSTRATE ****
52... $
53... ELEC      NUM=1     IX.L=36   IX.H=37   IY.L=4    IY.H=4
54... ELEC      NUM=2     IX.L=4    IX.H=33   IY.L=1    IY.H=1
55... ELEC      NUM=3     IX.L=1    IX.H=2    IY.L=4    IY.H=4
56... ELEC      NUM=4     IX.L=1    IX.H=37   IY.L=33   IY.H=33
57... $
58... $***define the doping profile ***
59... $
60... dop       new.sup boron     inf=n.exp    outf=n4.dop
61... dop       new.sup arsenic   inf=n+.exp   x.l=0.0 x.r=0.7   ratio=0.75
62... dop       new.sup pho       inf=n+.exp   x.l=0.0 x.r=0.7   ratio=0.75
63... dop       new.sup arsenic   inf=n+.exp   x.l=2.3 x.r=3.0   ratio=0.75
64... dop       new.sup pho       inf=n+.exp   x.l=2.3 x.r=3.0   ratio=0.75
65... $
66... plot.2d   boundary grid
67... $
68... end
```

Fig. 3.25. PISCES input for the n-channel MOSFET simulation.

Device Simulation

The first statement of all inputs should be a **title** statement. In **title** statement, heading should follow the **title**. The **comment** statement in PISCES-II starts with **$**. **Comment** statements can be placed anywhere. The program ignores whatever is written in **title** or **comment** statements. The **option** statement can also be placed anywhere. It specifies the various feature in PISCES simulation and postprocessing. In the input file, the option **x.s** specifies the x length of subsequent plots as 7 inches. This is necessary for hp2623 terminals. Convenient feature of PISCES-II input parameter name is that only the minimum number of characters to ensure unique identification are sufficient. For more details of input format and sequence, refer to overview (A.1) in PISCES-II manual [3.16].

A rectangular grid is specified by **mesh, x.mesh, y.mesh**, and **eliminate** statements. In the **mesh** statements, the number of grids in the x direction (**nx**) is 37 and in the y direction (**ny**) is 34. **Outfile** gives the name of a output mesh file as **n4msh3**.

```
$
X.M   N= 1    L=0.00  R=1.00
X.M   N= 6    L=0.81  R=0.70
X.M   N= 7    L=0.85  R=1.00
X.M   N=12    L=0.95  R=1.00
X.M   N=13    L=0.98  R=1.00
X.M   N=19    L=1.50  R=1.10
X.M   N=25    L=2.02  R=0.91
X.M   N=26    L=2.05  R=1.00
X.M   N=31    L=2.15  R=1.00
X.M   N=32    L=2.19  R=1.00
X.M   N=37    L=3.00  R=1.60
```

The actual position of x and y grids are specified by **x.mesh** and **y.mesh** statements. In both statements, **node** is the identification number of the grid point and **location** is the coordinate of that grid point in micron. **Ratio** is the ratio of the adjacent spacing in the previous grids. The spacing decreases with proportion to **ratio** when **ratio** is less than 1 and increases when **ratio** is greater than 1. The first **x.mesh** mesh statement defines the starting point. The **ratio** parameter in the first statement is always equal to one. The second statement allocated 5 uniform grids in the uniform source region. The third statement provides a transition from large spacing to small spacing near the source junction at 0.9 μm. The fourth statement puts 5 grid lines near the source junction. The fifth statement provides an intermediate spacing for the transition. The sixth and seventh statements allocate coarse grid in the channel region. The eighth statement is a transition line. The ninth statement allocates 5 fine grid lines near the drain junction. The tenth and eleventh statements are for the drain region.

```
$
Y.M   N= 1   L=-0.025     R=1.00
Y.M   N= 5   L=0.0000     R=1.00
Y.M   N= 6   L=0.0005     R=1.00
Y.M   N= 7   L=0.0010     R=1.00
Y.M   N= 8   L=0.0020     R=1.00
Y.M   N= 9   L=0.0040     R=1.00
Y.M   N=10   L=0.0080     R=1.00
Y.M   N=11   L=0.0160     R=1.00
Y.M   N=12   L=0.0320     R=1.00
Y.M   N=20   L=0.3000     R=1.10
Y.M   N=24   L=0.5000     R=1.10
Y.M   N=28   L=1.0000     R=1.10
Y.M   N=34   L=3.0000     R=1.20
```

The first and second **y.mesh** statements allocate 4 grids in the gate oxide region. To resolve the change of the vertical field and carrier concentration in the inversion layer, the first spacing in the inversion layer is 0.5 nm. The ratio of the adjacent spacing should be not greater than 2 to prevent numerical errors or convergence problems. Thus, the statements from the fourth to ninth increase the spacing only by 2 times. The tenth statement provides relatively fine grids for the depletion region below the channel. The next three statements gradually increase the spacing and allocate coarse grids in the substrate.

```
$ eliminate nodes to reduce the total node number
ELI   Y.DIR   IX.L=26    IX.H=32    IY.L=20    IY.H=34
ELI   Y.DIR   IX.L=07    IX.H=12    IY.L=20    IY.H=34
ELI   Y.DIR   IX.L=13    IX.H=16    IY.L=22    IY.H=34
ELI   Y.DIR   IX.L=22    IX.H=25    IY.L=22    IY.H=34
ELI   Y.DIR   IX.L=06    IX.H=32    IY.L=25    IY.H=34
ELI   Y.DIR   IX.L=01    IX.H=34    IY.L=29    IY.H=34
$
ELI   X.DIR   IX.L=01    IX.H=05    IY.L=06    IY.H=12
ELI   X.DIR   IX.L=33    IX.H=37    IY.L=06    IY.H=12
```

To reduce the number of nodes in the area where they are not needed, a **eliminate** statement will remove every other line in the specified area. The direction of grids to be eliminated is specified by **y.direction** or **x.direction**. The area to be eliminated is specified by four parameters (**ix.low, ix.high, iy.low, iy.high**). The first and second **eliminate** statements selectively remove the x grids (**Y.DIR**) below the source and drain junctions. The next four statements eliminate the x grids (**Y.DIR**) in the substrate. The last two statements remove the fine y grids (**X.DIR**) in the source and drain regions. These eliminations reduce the total number of nodes from 1258 to 939.

```
$
REGI  N=1    IX.L=1 IX.H=37        IY.L=1 IY.H=5
+     OXIDE
REGI  N=2    IX.L=1 IX.H=37        IY.L=5 IY.H=34
+     SILICON
$
$ ELECTRODES: 1-DRAIN 2-GATE 3-SOURCE 4-SUBSTRATE
$
ELEC  N=1    IX.L=36      IX.H=37     IY.L=05    IY.H=05
ELEC  N=2    IX.L=04      IX.H=33     IY.L=01    IY.L=01
ELEC  N=3    IX.L=01      IX.H=02     IY.L=05    IY.L=05
ELEC  N=4    IX.L=01      IX.H=37     IY.L=34    IY.H=34
```

The material type of the sub-area is specified by the **region** statement. **Number** is the identification (ID) number of the region. The boundary of the region is specified by **ix.low, ix.high, iy.low,** and **iy.high**. The first **region** statement specifies the gate oxide and the second statement specifies the silicon substrate. Next, **electrode** statements specify the contacts. In PISCES-II, a contact is a rectangular region specified by **ix.low, ix.high, iy.low,** and **iy.high**. The contact can collapse to a line if **ix.low=ix.high** or **iy.low=iy.high**. It can be located inside as well as on the boundary. Here, drain, gate, source, and substrate are specified as the first, second, third, and fourth electrodes. In the subsequent input, these contacts are referred by their ID numbers.

To specify the doping profiles in the device, the next statements should be a set of **doping** statements.

```
$ define the doping profile
doping     new.sup boron   inf=n.exp      outf=n4.dop
doping     new.sup ars     inf=n+.exp     x.l=0.0  x.r=0.7  ratio=0.75
doping     new.sup pho     inf=n+.exp     x.l=0.0  x.r=0.7  ratio=0.75
doping     new.sup ars     inf=n+.exp     x.l=2.3  x.r=3.0  ratio=0.75
doping     new.sup pho     inf=n+.exp     x.l=2.3  x.r=3.0  ratio=0.75
```

Device Simulation

The first **doping** statement specifies that the boron channel doping profile was created by a newer version of SUPREM-III (**new.sup**). The logical parameter **boron** specifies the dopant to be extracted from the SUPREM-III file. The SUPREM-III file name (**inf**) is given as **n.exp**. In the first **doping** statement, the implant window (**x.left, x.right**) is not specified. Thus, it is the uniform blank implant. The channel doping profile is shown in Fig. 3.26(a). The peak channel implant is 5×10^{16} cm^{-3} and the substrate doping is 8×10^{14} cm^{-3}. The second and third **doping** statements specify the double diffused source doping profile. The second statement tells the program to extract the arsenic profile from the SUPREM-III file (**n+.exp**) and the uniform-implant window is from 0.0 (**x.left**) to 0.7 μm (**x.right**). In this window, the doping profile is uniform in the x direction and only varies with y according to the 1-D SUPREM-III profile. Outside this window, the default lateral profile is the same shape as the vertical profile but it is scaled. The scale ratio (**ratio**) is 0.75. The third statement specifies the same information except the dopant is phosphorus. The fourth and fifth **doping** statements specify the double diffused drain doping profile. All the specifications are the same as those of the source except the implant window is from 2.3 μm to 3.0 μm. The specified doping profiles yield an effective channel length of 1.2 μm. The vertical doping profile of the source and drain regions is shown in Fig. 3.26(b). Due to the arsenic and phosphorus double diffusion, the profile is smoother than the conventional arsenic-only profile. The junction depth is about 0.27 μm. The **end** statement marks the end of the input file. The device structure specified by this input is shown in Fig. 3.27(a). The final mesh saved in **n4msh3** is shown in Fig. 3.27(b). The total number of nodes is 939.

Electrical characteristics simulation

The next phase is the actual device simulation in the subthreshold region. In this input, the first step is to load the device structure, a mesh and doping information from the mesh file **n4msh3** using a **mesh** statement. Before solving the equations, several things should be specified.

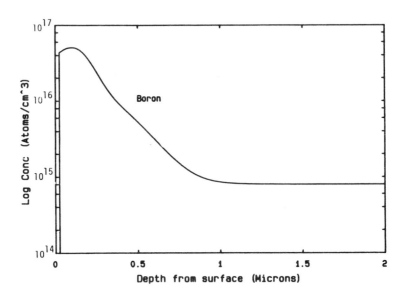

(a)

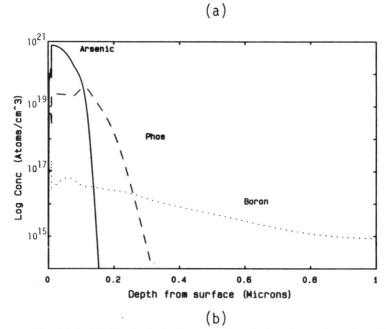

(b)

Fig. 3.26. (a) Vertical doping profile of the channel region. (b) Vertical doping profile of the source and drain regions.

Device Simulation

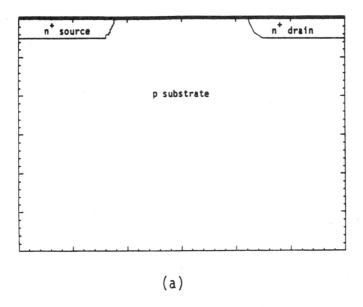

(a)

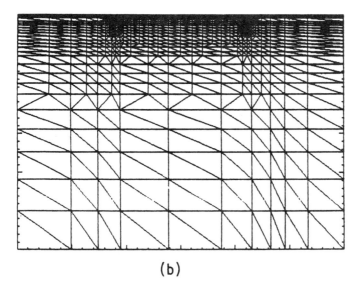

(b)

Fig. 3.27. (a) Device structure generated by the input. (b) Grid generated by the input.

```
title subthreshold & linear region characteristics : n4b.i
$
$ load the mesh file
mesh   infile=n4msh3
$
$ specify the symbolic factorization and parameters
symbolic      newton carriers=1      electrons
method        autonr
$
$ specify the contacts and modes
contact       num=2 n.poly
models        temp=300       print
```

In an n-channel MOS device, it is enough to solve the Poisson and electron continuity equations to calculate the channel current because the hole current is negligible. Thus, the number of carriers (**carriers**) is specified as 1 and the carrier type as electron (**electrons**) in the **symbolic** statement. In this input, the solution method is specified as the direct Newton method (**newton**) using an automated Newton-Richardson procedure (**autonr**) as shown in the **symbolic** and **method** statements.

The next thing is to specify the material types (or work function) of electrodes using the **contact** statement. The gate is the second electrode (**number=2**). It is n^+ polysilicon (**n.poly**). Other electrodes are not specified. Thus, neutral contacts will be assigned as defaults. The physical models to be used in the simulation should be specified such as the mobility model, the recombination model, or statistics, etc. These models are specified by a **models** statement. The most important physical model is the mobility model in the MOS device simulation. In the subthreshold region, however, it is not critical so that the default constant mobility model is used. The absolute temperature is also set to 300 ^{0}K in the **models** statement. The logical parameter **print** directs the program to print out all the information on the physical models used in the simulation. Next, the initial solution will be solved.

```
$ solve the initial bias point
solve   initial   outf=n4slv0
$
$ set up I-V characteristics log file
log     outfile=n4b.IV
$
$ solve for Vds=0.1
solve   v1=0.1
```

Since there has been no solution before, the solution should start from the initial guess (**initial**) where all biases on the electrodes are equal to zero. In the first **solve** statement, all biases are set to zero by default and the solution will be saved in the file, **n4slv0**. Before proceeding further, a log file for the biases and currents of all electrodes is set up using a **log** statement. This log information will be saved in the file, **n4b.IV**. It will be used to plot I-V characteristics later. The next **solve** statement increases the bias of the first electrode (drain) to 0.1 V and calculates the solution. The gate voltage is ready to be stepped to generate the subthreshold and linear region I-V characteristics.

```
$ step Vgs from 0 to 2.0 volts
solve   electrode=2    vstep=0.2      nsteps=9
solve   v2=2.0         outf=n4slv1
```

In the **solve** statement, the bias of any electrode can be set to a new constant value. If new values are not set, previous values will be used. The bias of the electrode can also be stepped with a uniform increment as shown on the third **solve** statement in the input file. Usually, the bias step should be less than 0.5 V. Otherwise, the convergence may be slow or non-convergence may occur in the worst case. In the bias stepping, the program needs the identification number of the electrode (**electrode**), the step voltage (**vstep**), and the number of steps (**nsteps**). In the third **solve** statement, the second electrode (gate) is stepped by 0.2 V from 0 V to 1.8 V. The method of initial guess for the solution can also be specified in the **solve** statement.

The first bias point for a given structure must have the **initial** parameter specified. From then, the program will either use the previous (**previous**), or if there are two previous solution present and equivalent bias steps are taken on any electrodes that are changed, an extrapolation (**project**) from the preceding two solutions will be used to get an improved initial guess. After the initial bias point, the program will automatically use extrapolation wherever possible if no initial guess parameter is supplied. The extrapolated initial guess may reduce the number of iterations especially for bias stepping. The biases of the multiple electrodes can be stepped at the same time. If more than one electrode is to be stepped, **electrode** should be an n-digit integer, where each of the n-digits is a separate electrode number. If only the last bias solution is to be saved, the output file name should not be specified in the solve statement with bias stepping. Otherwise, the program will save each solution in separate files. Use a separate **solve** statement for the last bias point and specify the output file name as shown in the fourth **solve** statement. In summary, the four **solve** statements solve for the subthreshold and linear region characteristics of the n-channel MOS device for the drain bias of 0.1 V and the gate bias from 0 V to 2.0 V.

The total input for the subthreshold characteristics is shown in Fig. 3.28(a). A typical output for a bias point is also illustrated in Fig. 3.28(b). For each bias point, the iteration information is first printed. The default termination criteria are that the maximum potential change is less than 1×10^{-5} kT/q for the Poisson equation and that the maximum change relative to the local concentration is less than 10^{-5} for the current continuity equations. Both criteria must be satisfied. These criteria can be changed with the **p.tol** and **c.tol** parameters in the **method** statement. After convergence, the program calculates the conduction, displacement and total currents per micron width and prints out the results. For a DC simulation, the displacement current is zero and the conduction and total currents are the same. The cpu time is also printed out in seconds.

Device Simulation

```
title    CMOS Subthreshold Characteristics - n4b.i
$
$        Load the mesh
mesh     infile=n4msh3
$
$        Indicate the symbolic factorization and parameters
symb     newton carriers=1 electrons
method   autonr
$
$        Indicate the materials and contacts
contac   num=1 n.poly
$
$        Indicate the models
models   temp=300  print fldmob
$
$
$        Solve the initial bias point
solve    initial  outf=n4slv0
$
$        Setup IV log file
log      outfile=n4b.IV
$
$        Solve for Vds=0.1
solve    v4=0.1
$
$        Step Vgs from 0 to 2.0 volts
solve    electrode=1 vstep=0.2 nsteps=9
solve    v1=2.0 outf=n4slv1
$
end
```

(a)

```
Solution for bias:
  V1 =  1.0000000E-01      V2 =  2.0000000E+00
  V3 =  0.0000000E+00      V4 =  0.0000000E+00

Projection used to find initial guess

iter   psi-error    n-error      p-error
  1    5.0516E-02   6.3364E-02   0.0000E+00
  2    1.2486E-03   6.0684E-04   0.0000E+00
  3*   2.4226E-08   4.1971E-07   0.0000E+00

Electrode  Voltage   Electron Current  Hole Current     Conduction Current
           (Volts)   (amps/micron)     (amps/micron)    (amps/micron)
    1       .1000     1.25829E-05      -0.00000E+00      1.25829E-05
    2      2.0000    -0.00000E+00      -0.00000E+00     -0.00000E+00
    3       .0000    -1.25829E-05      -0.00000E+00     -1.25829E-05
    4       .0000    -5.73607E-17      -0.00000E+00     -5.73607E-17

Electrode       Flux          Displacement Current    Total Current
            (coul/micron)        (amps/micron)        (amps/micron)
    1        2.41950E-20         0.00000E+00          1.25829E-05
    2        4.68779E-15         0.00000E+00          0.00000E+00
    3       -2.10775E-20         0.00000E+00         -1.25829E-05
    4        6.41391E-22         0.00000E+00         -5.73607E-17

Absolute convergence criterion met for Poisson
Absolute convergence criterion met for continuity
Total cpu time for bias point =  281.35
Total cpu time = 4865.18
Solution written to n4slv1
```

(b)

Fig. 3.28. (a) Input for the subthreshold simulation. (b) A typical output for a bias point in PISCES.

Graphical post-processing

The initial solution is saved in **n4slv0**. The last solution is saved in **n4slv1**. The I-V log information is saved in **n4b.IV**. The I-V characteristics are plotted using the following input file.

```
title plot I-V : n4piv.i
$
option   x.s = 7
$
mesh     inf = n4msh3
$
$ plot Id vs Vgs  (log scale)
plot.1d  inf = n4b.IV x.axis = v2 y.axis = i1 abs log pause
$
$ plot Id vs Vgs  (linear scale)
plot.1d  inf = n4b.IV x.axis = v2 y.axis = i1
end
```

Although the mesh information is not used, the mesh file should be read first because the program prints an error message otherwise. The I-V information is read back from **n4b.IV** in the **plot.1d** statement. The x axis (**x.axis**) is assigned to V_G (**v2**) and y axis (**y.axis**) to I_D (**i1**). As a default, the plot ranges of x and y variables are determined by the maximum and minimum values of x and y variables in the input file. **Absolute** and **logarithm** direct the program to take the absolute value and logarithm of the y variable. The I-V plot generated by this statement is shown in Fig. 3.29(a). From this plot, the subthreshold slope can be determined. The subthreshold slope, S is 98 mV/decade by simulation, which is close enough to the measured value of 96 mV/decade. After this plot, the program stops and waits because the **pause** parameter is set. One can examine the plot or make a hard copy and continue the program by pushing the return key. The second **plot.1d** line plots the I-V characteristics on a linear scale. This plot is shown in Fig. 3.29(b). The extrapolated V_T is 0.8 V for this short-channel device, which is comparable to the measured value of

Device Simulation

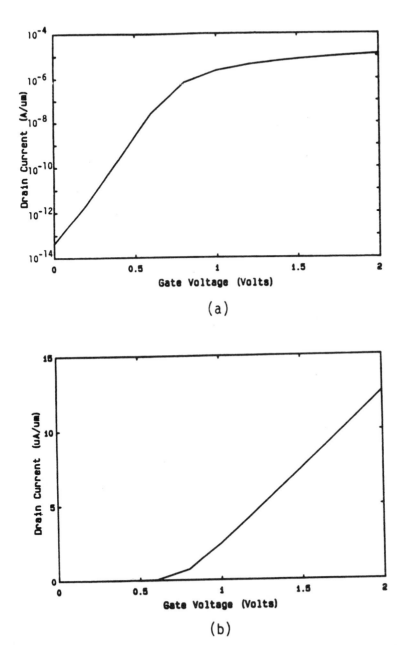

Fig. 3.29. (a) Logarithmic plot of I_D vs V_G. (b) Linear plot of I_D vs V_G.

0.75 ± 0.04 V.

The internal distributions of major variables can be plotted using **plot.2d** and **contour** statement. The input for the potential contour plot is as follows.

```
title plot 2-D contour plot of potential : n4p2.i
$
opt          x.s=7
$
$ load the mesh and solution files
mesh         infile=n4msh3
load         infile=n4slv1
plot.2d      boundary      junction depletion
contour      potential     min=-1 max=2 del=0.1
end
```

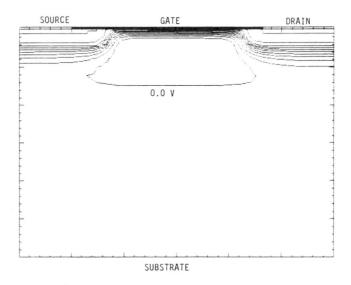

Fig. 3.30. Contour plot of potential when $V_G = 2$ V and $V_D = 0.1$ V. Increment of Contour is 0.1 V.

Device Simulation 123

First, the mesh file is read using a **mesh** statement. The solution file is read using a **load** statement. Flags for the plots of the boundary, metallurgical junction, and depletion boundary are set in the **plot.2d** statement. In the **contour** statement, the variable for the contour plot is given as **potential**. The minimum (**min**), maximum (**max**), and increment (**del**) of the contour are also specified. The resulting contour plot is shown in Fig. 3.30. The contour plot gives a good overview of the 2-D nature of the distribution.

The input of 1-D cross-sectional plot of doping along and vertical to the channel is given below.

```
title plot 1-D cross-sectional plot of doping : n4p1.i
$
option      x.s=7
$
$ load the mesh and solution files
mesh  infile=n4msh3
load  infile=n4slv1
plot.1d       doping abs    log    min=10 max=20
+             x.s=0.0 y.s=0.0 x.e=3.0 y.e=0.0 pause
plot.1d       doping abs    log    min=10 max=20
+             x.s=1.5 y.s=0.0 x.e=1.5 y.e=1.0
end
```

The mesh and solution files are read first as in the previous example. In the **plot.1d** statement, doping (**doping**) is specified as a plot variable and **absolute** and **logarithm** parameters are set to plot the logarithm of the absolute value of doping density. Thus, the ranges of doping (**min, max**) are given from 10 (10^{10} cm^{-3}) to 20 (10^{20} cm^{-3}). For a log plot, **min** and **max** are the value of the base 10 logarithm. There are two plots in the example. In the first 1-D plot, The cross-section statement is along the surface channel. The coordinates of the starting point (**x.s, y.s**) is (0.0, 0.0) and that of the ending point (**x.e, y.e**) is (3.0, 0.0). As mentioned before, the origin of the coordinate system is at the

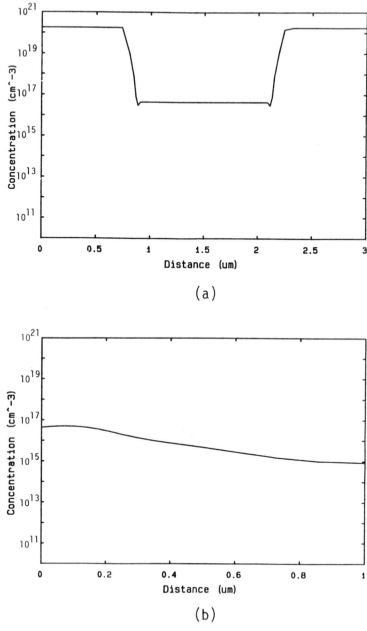

Fig. 3.31. (a) 1-D plot of doping profile along the surface channel. (b) 1-D plot of vertical doping profile in the middle of the channel.

Device Simulation

upper left corner of substrate. The positive x direction is from left to right. The positive y direction is from top to bottom. The cross-sectional plot of doping along the surface channel is shown in Fig. 3.31(a). The second plot is along a vertical slice in the middle of the channel. The coordinates of the starting and ending points are (1.5, 0.0) and (1.5, 1.0). The plot is shown in Fig. 3.31(b).

As shown above, the graphical post processing is very important. It should be used to check the simulation structure to ensure correct dimensions and doping profiles. Second, useful device parameters such as a threshold voltage can be extracted from the I-V characteristics. Third, plots of internal distributions will help to understand the device operation and terminal characteristics. In the above example, the plot of doping profile is usually done to check that the transfer of doping from the SUPREM-III files is correct. Other usages of the graphical post-processing will be given in the second part of this book.

References

[3.1] J. A. Greenfield and R. W. Dutton, "Nonplanar VLSI Device Analysis Using the Solution of Poisson's Equation," *IEEE Trans. on Electron Devices*, **ED-27**, Aug 1980, pp. 1520-1532.

[3.2] S. Ogura, P. J. Tsang, W. W. Walker, D. L. Chritchlow, and J. F. Shepard, "Design Characteristics of the Lightly Doped Drain-Source (LDD) IGFET," *IEEE Trans. on Electron Devices*, **ED-27**, Aug 1980, pp. 1359-1367.

[3.3] R. D. Rung, H. Momose, Y. Nagakubo, "Deep Trench Isolated CMOS Devices," *Tech. Digest of IEDM 1982*, pp. 237-240.

[3.4] K. M. Cham, S. Y. Chiang, D. Wenocur, and R. D. Rung, "Characterization and Modeling of the Trench Surface Inversion Problem for the Trench Isolated CMOS Technology," *Tech. Digest of IEDM 1983*, pp. 23-26.

[3.5] R. S. Verga, *Matrix Iterative Analysis*, Englewood Cliffs, NJ:Prentice-Hall, 1962, ch.6.

[3.6] J.M. Ortega and W. C. Rheinboldt, *Iterative Solution of Nonlinear Equation in Several Variables*, New York:Academic Press, 1970, pp. 214-230.

[3.7] T. Toyabe, *"CADDET User's Manual"*, Hitachi Central Laboratories

[3.8] M. S. Mock, *"Analysis of Mathematical Models of Semiconductor Devices,"* Boole Press, Dublin, 1983.

[3.9] D. Vandorpe, J. Borel, G. Merkel, and P. Saintot, "Accurate Two-Dimensional Numerical Analysis of the MOS Transistor," *Solid State Electronics*, 1972, **Vol. 15**, PP. 547-557.

[3.10] H. L. Stone, "Iterative Solution of Implicit Approximations of Multidimensional Partial Difference Equations," *SIAM J. Numerical Anal.*, **5**, 1968, pp. 530-558.

[3.11] D. L. Scharfetter and H. K. Gummel, "Large-signal Analysis of a Silicon Reed Diode Oscillator," *IEEE Trans. on Electron Devices*, **ED-16**, pp. 64-77, Jan 1969.

[3.12] K. Yamaguchi, "Field-Dependent Mobility Model for Two-Dimensional Numerical Analysis of MOSFETs," *IEEE Trans on Electron Devices*, **Vol ED-26**, pp. 1068-1074, July 1978.

[3.13] K. Yamaguchi, " A Mobility Model for Carriers in the MOS Inversion Layers," *IEEE Trans. on Electron Devices*, **Vol Ed-30**, pp. 658-663, June 1983.

[3.14] S. Y. Oh, P. Vande Voorde, and J. Moll, "An Empirical Mobility Model for Numerical MOSFET Simulation," *Hewlett-Packard Semiconductor Technology Conference Proc.*, pp. 97-104, 1984.

[3.15] K. K. Thornber, "Relation of Drift Velocity to Low-Field Mobility and High Field Saturation Velocity," *J. Appl. Phys.*, Vol. 51, No. 4, pp. 2127-2136, April 1980.

[3.16] M. R. Pinto, C. S. Rafferty, R. W. Dutton, " PISCES II: Poisson and Continuity Equation Solver," Stanford Electronics Laboratory, Stanford

University, California, Sept., 1984.

[3.17] C. H. Price, "Two-Dimensional Numerical Simulation of Semiconductor Devices, " Stanford Electronics Laboratory, Stanford University, California, May, 1982.

Chapter 4
Parasitic Elements Simulation

4.1 Introduction

As we scale down the critical dimensions in integrated circuits, the effect of interconnects becomes as critical as that of devices on the overall circuit performance [4.1],[4.2]. The interconnect lines not only act as loads for their drivers but also become a source of noise because the lines are capacitively coupled when they are close to each other. Also, because we scale down the widths of the lines while the lengths of the lines are generally fixed, the resistance of the lines becomes larger. Generally, parasitic inductance is not of concern in on-chip interconnections. However, it is very critical in the packaging of integrated circuits. Careful characterization of these parasitic components in integrated circuits is essential to improve the performance of the circuits.

In this chapter, we discuss simulation programs to extract the parasitic components in integrated circuits. For many of the interconnection geometries where the lines are parallel, two-dimensional simulations give good approximations. However, for geometries like crossing lines and via contact in a multi-level interconnect system, three-dimensional calculations are necessary. To solve the above problems, two-dimensional and three-dimensional simulation programs, SCAP2 and FCAP3, are presented in this chapter. We focus on the numerical techniques used in the programs and some basic application examples of the programs. Chapter 15 gives more extensive application examples.

4.2 SCAP2 : Two-Dimensional Poisson Equation Solver

SCAP2 is a two-dimensional Poisson equation solving program. For a given set of geometries and bias condition, SCAP2 solves the Poisson equation using the finite difference method with self-adjusting rectangular grid and Incomplete Cholesky Conjugate Gradient (ICCG) method [4.3].

A user can input two-dimensional geometries using the elements shown in Fig. 4.1 which were developed in FCAP2 [4.4] or can build structures using arbitrary polygons. The polygonal input scheme was developed in order to link the two-dimensional process simulator, SAMPLE [4.5], to the FCAP2. (Hence, the name of the program was changed from FCAP2 to SCAP2 (SAMPLE + FCAP2)). The user can simulate a problem very easily with these geometrical elements of FCAP2 or the polygonal inputs.

When all the geometries are defined, SCAP2 generates a semi-uniform grid based on the vertices of the geometries. According to the dielectric constants of the insulators and the potential values on the conductors of the problem, SCAP2 sets up a matrix using the five-point difference equation over the grid. The reflective (Neumann) boundary condition is applied on the boundaries of the simulation window. The user can exploit this boundary condition for a symmetric or periodic structure to minimize the size of the simulation window. However, when the geometry of a problem is isolated, the simulation window should be large enough to reduce the effect of the mirror images outside of the simulation window. The matrix is solved by the ICCG method which provides a rapid convergence to the solution. From the calculated potential distribution, SCAP2 extracts the amount of charge on each conductor for the given bias condition. The user can obtain the capacitance matrix of a multi-conductor system by applying different sets of biases on the conductors.

One of the good features of SCAP2 is its automatic regrid capability. Since potential varies more rapidly in some areas than others in most of the

Parasitic Elements Simulation

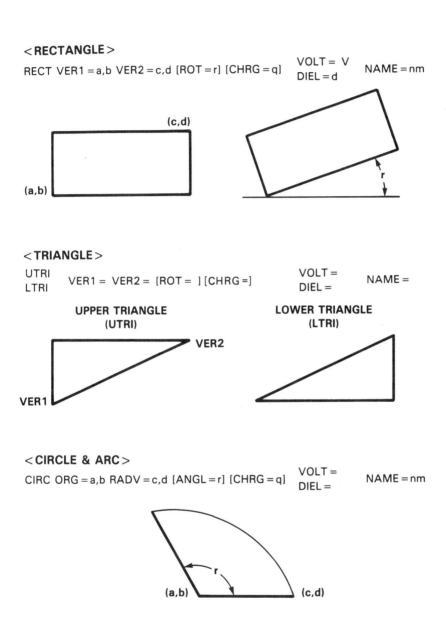

Fig. 4.1. Geometry input elements of SCAP2.

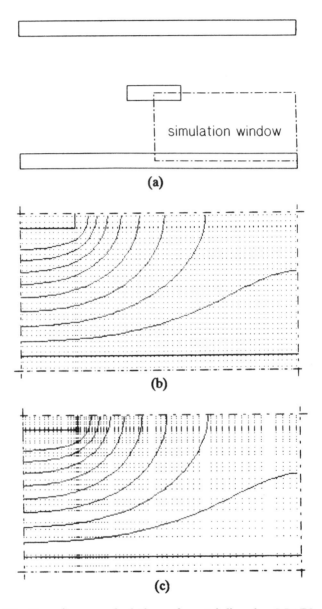

Fig. 4.2. Capacitance calculation of a stripline by SCAP2. (a) Geometry of a stripline. (b) Potential distribution on the initial semi-uniform grid. (c) Potential distribution on the automatically adjusted grid.

problems, it is desirable to have the denser grid in the area of greater potential gradient. Thus, the initial semi-uniform grid is not suitable for the problem. However, since the initial calculation gives the rough potential distribution over the solution area, we can generate a new grid based on the potential distribution. When the regrid command is given after the initial iteration, SCAP2 generates a new grid which conforms with the potential variation. The next iteration on this new grid gives a more accurate solution. By repeating this regrid and iteration, SCAP2 generates the optimal grid for the problem, hence, giving the best solution.

Fig. 4.2(a), (b), and (c) show the capacitance calculation of a strip line. The two-fold symmetry allows us to simulate only one quarter of the problem. The geometries and potential contour plot over the initial semi-regular grid are shown in Fig. 4.2(b). A new grid based on the previous potential calculation is shown in Fig. 4.2(c). Note that we have denser grid points under the signal line where the potential gradient is larger than the rest of the area.

The characteristic impedance of the strip line in Fig. 4.2(a) can be easily calculated with SCAP2 for TEM wave propagation with quasi-static approximation. By doing the same calculation as before except for removing the dielectric, we can find the capacitance, C_0. The inductance at the high frequency limit, L, can be obtained by

$$L = \frac{1}{c^2 C_0}. \qquad (4.1)$$

where c is the speed of light in vacuum. The characteristic impedance of the strip line, Z_c, is

$$Z_c = \left(\frac{L}{C}\right)^{1/2} \qquad (4.2)$$

where C is the capacitance with the dielectric which was calculated previously.

The capacitance calculation of polygonal conductors is shown in Fig. 4.3(a), (b), and (c). There are more grid points between two bean-shaped conductors than the other area on the reallocated grid in Fig. 4.3(c). Consequently, the shapes of the inner sides of the conductors are described better than their outer side. This new grid gives a more accurate capacitance value than the initial grid in Fig. 4.3(b) because most of the charges are located at

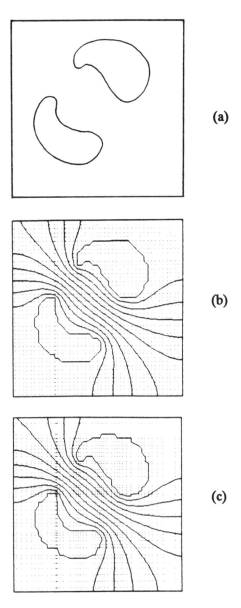

Fig. 4.3. Capacitance calculation of polygonal geometries by SCAP2. (a) Two bean-shaped conductors. (b) Potential distribution on the initial semi-uniform grid. (c) Potential distribution on the automatically adjusted grid.

Parasitic Elements Simulation

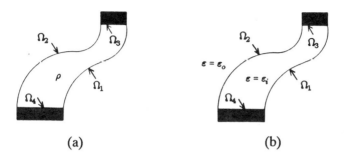

Fig. 4.4. Duality between capacitance and resistance structures.

the surface of the inner sides of the beans.

SCAP2 can calculate the resistance of a two-dimensional geometry as well as its capacitance using the duality between them. A geometry of a resistive material whose resistance we want to calculate is shown in Fig. 4.4(a). Let us assume that the geometry has a unit depth into the paper. Intuitively, the field lines in the resistor would be the same as those in the capacitor shown in Fig. 4.4(b) if the ratio of the dielectric constant of the capacitor to the dielectric constant of the ambient were infinite. This is theoretically true because we solve the Laplace equation with the same boundary conditions inside the resistor and the capacitor if the assumption for the dielectrics is true. The boundary conditions for the resistor are

$$\nabla \phi \cdot \mathbf{n} = 0 \quad \text{on} \quad \Omega_1, \Omega_2 \qquad (4.3)$$

$$\nabla \phi \times \mathbf{n} = 0 \quad \text{on} \quad \Omega_3, \Omega_4 \qquad (4.4)$$

and for the capacitor

$$\varepsilon_i \nabla \phi \cdot \mathbf{n} = \varepsilon_0 \nabla \phi \cdot \mathbf{n} \quad \text{on} \quad \Omega_1, \Omega_2 \qquad (4.5)$$

$$\nabla \phi \times \mathbf{n} = 0 \quad \text{on} \quad \Omega_3, \Omega_4 \qquad (4.6)$$

where ϕ, ε_i, ε_0, and $\mathbf{n}$ are potential, dielectric constant of the capacitor, dielectric constant of the vacuum, and a unit normal vector toward outside, respectively. The boundary conditions for resistance and capacitance become identical when $\varepsilon_0/\varepsilon_i$ becomes 0. To be practical, we set the relative dielectric constant of the capacitor dielectric to be 1 and that of the ambient to be 10^{-30} or smaller. With this condition, we can find the value of the capacitor with almost no field line leaking out of the dielectric. Then we can easily convert the capacitance, C, to the resistance, R, of the same shape in Fig. 4.4(a) using

$$R = \frac{\rho \varepsilon_i}{C} \quad (4.7)$$

where ρ and ε_i are resistivity and dielectric constant, respectively. This method to find a resistance value from the capacitance of the same shape extends itself to three-dimensional cases as discussed in the next section.

4.3 FCAP3 : Three-Dimensional Poisson Equation Solver

As discussed in section 4.1, there are cases where three-dimensional simulation is necessary to calculate the parasitic components of the structures. FCAP3 (Fast CAPacitance 3-dimensional program) [4.6] is a Poisson equation solver in three-dimensional space which was developed to attack these kinds of problems. FCAP3 is basically an extension of SCAP2 from two-dimensional space to three-dimensional space. However, the extension is not trivial as readers may expect. In this section, we discuss the algorithm, implementation of user-friendliness, and basic applications of FCAP3.

FCAP3 solves the three-dimensional Poisson equation using the finite-difference method and ICCG method like SCAP2. When a problem is given, FCAP3 generates a semi-uniform rectangular grid with 125,000 grid points. The grid lines are initially allocated at the vertices of the geometries and the rest of grid lines are allocated evenly over the simulation region. After geometries are discretized with the grid, a matrix is set up using the seven-point finite-difference equation with the reflective (Neumann) boundary condition applied on the simulation boundary planes. Then the potential distribution of the problem is obtained by solving the matrix using the ICCG method. FCAP3 is capable of reallocating its grid based on the initial potential distribution. The new grid discretizes the geometries more appropriately since there are more grid lines where the potential gradient is large. The description of a geometry or a part of a geometry is poor if the potential gradient in the region is small because the region is discretized with a sparse grid. However, this is

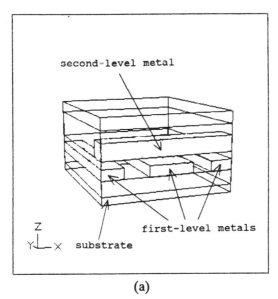

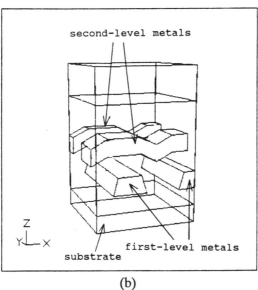

Fig. 4.5. Multi-level on-chip interconnects simulated by FCAP3. (a) A drawing from the first-level FCAP3 input. (b) A drawing from the second-level FCAP3 input.

quite reasonable since the region does not have large effect on the amount of charge on the conductors which we want to calculate. This automatic regrid capability of FCAP3 makes the finite-difference method quite useful for three-dimensional problems, even though the finite-difference method is generally inferior to the finite-element method in describing complex geometries. Also, it generates a new grid without any user interaction, whereas a finite element grid is very difficult to automatically reallocate or even allocate, especially in three-dimensional space.

User-friendliness is a very important factor in three-dimensional simulation programs since it is difficult to input the geometries of a problem and to interpret the output of the problem. FCAP3 provides three degrees of complexity in input geometries to make input easy. The first level input deals with box shape geometries which are parallel to the axes of the rectangular coordinates. The second level input includes parallelograms which are parallel to the y-axis and have arbitrary polygonal side walls parallel to the xz-plane. The first and second levels are mainly for the typical geometries which occur in integrated circuits. Examples of two-level interconnect lines implemented with the first and the second level geometries are shown in Fig. 4.5(a) and (b), respectively. The third level input is open to users. Users can build their own geometry defining routines according to their needs. The following example of capacitance calculation of concentric sphere is built using the third level input scheme.

The calculation of the capacitance of concentric spheres, for which the finite difference method is the least adequate, supports the argument on the merit of the finite-difference method with the automatic regrid and verifies the accuracy of FCAP3. Fig. 4.6(a) shows an agreement within 9 % between the theoretical value (solid line) and FCAP3 results (circles). The optimal finite-difference grid generated by the automatic regrid is shown in Fig. 4.6(b). The grid was taken on the plane which goes through the center of the spheres and parallel to the xz-plane. The potential contour plot on the plane in Fig. 4.6(c) shows almost circular rings. This simulation result demonstrates the accuracy and versatility of FCAP3. An accuracy verification against an experiment is given in section Chapter 15.

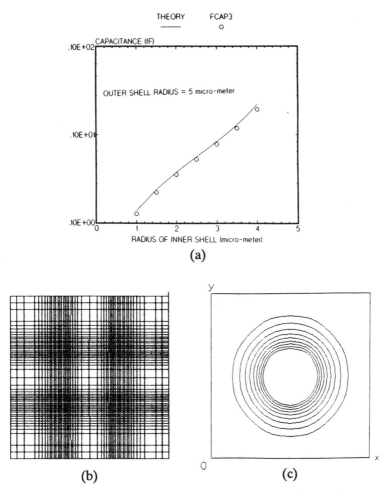

Fig. 4.6. Capacitance calculation of two concentric conductor spheres by FCAP3. (a) Comparison between theory and FCAP3 results. (b) Automatically generated grid based upon the initial calculation with a uniform grid. (c) Equipotential contours on a plane which goes through the center.

FCAP3 can also be used for the resistance calculation of complex three-dimensional resistive paths. The same argument in the previous section for the SCAP2 applies when it is extended to three-dimensional space. The same conversion equation (4.7) is valid for three-dimensional cases. An example of three-dimensional resistance calculation is given in section 15.3.

References

[4.1] Special Issue on Interconnections and Contacts for VLSI, *IEEE Trans. Electron Devices*, **ED-34**, no.3, Mar. 1987.

[4.2] K. Lee, Y. Sakai, and A.R. Neureuther, "Topography Dependent Electrical Parameter Simulation for VLSI Design," *IEEE Trans. Electron Devices*, **ED-30**, no.11, pp.1469-1474, Nov. 1983.

[4.3] D. S. Kershaw, "The Incomplete Cholesky - Conjugate Gradient Method for the Iterative Solution of Systems of Linear Equations," *Journal of Computational Physics 26*, pp.43-65, 1978.

[4.4] HP Internal Software.

[4.5] W. G. Oldham, A. R. Neureuther, C. Sung, J. L Reynolds, and S. N. Nandgaonkar, "A General Simulator for VLSI Lithography and Etching Processes : Part II - Application to Deposition and Etching," *IEEE Trans. Electron Devices*, **ED-27**, no. 8, pp. 1455-1459, Aug. 1980.

[4.6] HP Internal Software.

PART B

Applications and Case Studies

Chapter 5

Methodology in Computer-Aided Design for Process and Device Development

The previous chapters have presented an overview of computer-aided design (CAD) in VLSI development, as well as the simulation tools currently used at Hewlett-Packard Laboratories. In this chapter, CAD is discussed from the user point of view. The methodology for using the simulation tools in the most effective way is presented. Then case studies will be presented in the following chapters which show in detail how simulation tools are used in device designs.

5.1 Methodologies in Device Simulations

Simulation tools should be used in the most efficient way, such that time and effort in doing the simulations are minimized. This is especially true in process development where time is of major concern. The goal is to provide process parameters in the shortest time. Ideally, one would like to have CAD tools which can produce the desired output in minimal time. In real life, this is usually not possible, mostly because of limitations in software and hardware capabilities, or due to too many users on the system. In the following paragraphs, we discuss the simulation methodologies which will use the CAD tools most effectively:

1) Before simulations are to be performed, it is always a good idea to look at the problem from the simplest and most intuitive point of view, and to try to get some basic idea of the problem. For example, in the case of counter-doping for the p-channel transistors with n^+ polysilicon gate (to be discussed in detail in Chapter 12), the dose can be roughly estimated by simple arguments, and it turns out to be not very different from the results of two-dimensional analysis. This simple procedure not only provides a first estimate of the magnitude of the process parameters to be used, but also provides a simple picture of the physics behind the technique. Numerical analysis is sometimes difficult to interpret unless one has a basic idea of the physics involved. Also by gaining some initial knowledge of the problem, the simulation work can be better planned and executed, with a minimum range of parameter values, instead of using trial and error.

2) A simulation should be considered as an experiment itself. This means that a systematic procedure should be used rather than shooting for a particular number. For example, it may be desired to develop a p-channel MOSFET with a particular long channel threshold voltage, say, -0.7 V, and the threshold voltage is dependent on the counter-doping of boron [5.1]. In the simulation work, a factorial experiment [5.2] should be set up with the boron implant dose and energy as parameters. The simulated results are shown in Fig. 5.1. This figure provides an overall picture of the dependence of the threshold voltage on the two parameters. Also it provides an idea of the reasonable range of the parameters that should be used in the fabrication experiment. To be more complete, the n-well impurity surface concentration should be included in the simulation. In this case, the simulations become a 2^3 factorial experiment. The results, when arranged in a 3-dimensional form as shown in Fig. 5.2, show the dependence of the threshold voltage on the three parameters. The threshold voltage change due to the change in a combination of the parameters can be estimated quickly from the figure. Several observations can be made from this simple example. Fig. 5.3 shows the threshold voltage (V_T) as a function of the counter-

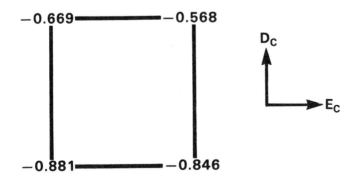

Fig. 5.1. P-channel threshold voltage vs. counter-doping energy (E_C) and dose (D_C).

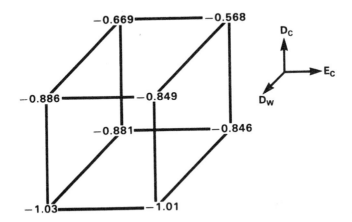

Fig. 5.2. P-channel threshold voltage vs. counter-doping energy (E_C), dose (D_C) and n-well implant dose (D_W).

SUPREM SIMULATIONS

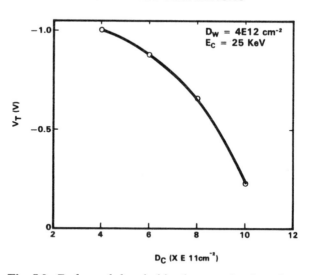

Fig. 5.3. P-channel threshold voltage vs. implant dose D_C.

doping dose (D_C) over a wide range. The dependence is nonlinear over the whole range, but can be approximated to be linear within a smaller range. (Nonlinearity is most serious for the counter-doping case. For n-channel device, the threshold versus channel implant dose are much more linear in a larger range of doses.) Fig. 5.4 shows the dependence of the threshold voltage on the counter-doping implant energy (E_C) for two different doses and a fixed n-well implant dose. Fig. 5.5 shows the dependence of the threshold voltage on the counter-doping dose for two different n-well implant doses (D_W) and a fixed counter-doping implant energy. The results show that the dependence of the threshold voltage on the different parameters are correlated, as can be observed by the different slopes of the curves in each figure. This means that for a narrow range of the parameters, the threshold voltage can be expressed as:

$$V_T = a_1 D_C + a_2 E_C + a_3 D_W \\ + b_{11} D_C E_C + b_{12} D_C D_W + c D_C E_C D_W \qquad (5.1)$$

Methodology in CAD

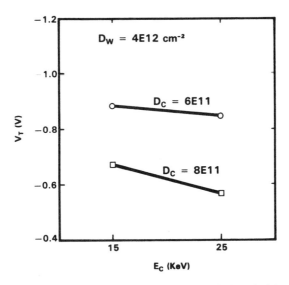

Fig. 5.4. SUPREM simulation of p-channel threshold voltage vs. counter-doping energy for two different doses.

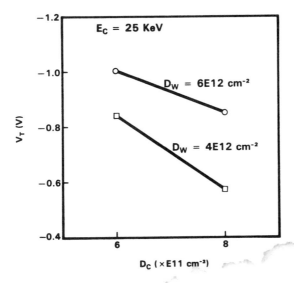

Fig. 5.5. SUPREM simulation of p-channel threshold voltage vs. counter-doping dose for two different n-well implant doses.

In most cases, however, the coupling coefficients b_{ij} and c are smaller than the coefficients a_i, hence making the problem much simpler. As a first approximation, the parameters D_C, E_C, and D_W can be assumed to be linearly independent variables. The dependence of the threshold voltage can thus be expressed as:

$$V_T \approx a_1 D_C + a_2 E_C + a_3 D_W \qquad (5.2)$$

In this case, four calculations only will be needed to determine the three coefficients a_i's, while it will take seven calculations to solve for all coefficients in Eq. (5.1). Therefore, there is significant savings in CPU time. The saving becomes more significant as the number of process parameters increases.

3) In most cases, and especially in the early stages of experiments, the trend rather than exact values generated by simulations should be emphasized. For example, in the above mentioned experiment on the threshold voltage, it would be a bad idea to try to generate the dose and energy that would give a threshold value of exactly -0.7 V. There are two reasons for this; first, the simulation tools are not always very accurate. The accuracy depends on the particular structure being considered and the simulation tools being used. Second, there are process control variations during fabrication, such as the line width and dielectric thickness variations. In VLSI, device performance is very sensitive to structural dimensions, so it is very important to understand the sensitivities. The simulations provide an optimized range of parameters for experimentation, which then provides feedback on the accuracy of the simulations. Through this feedback loop, the optimized process can be developed.

4) The simplest approach should be used in the simulations. It is a waste of time to simulate a structure with more accuracy than is necessary. For example, if one wants to calculate the threshold voltage of a long channel MOSFET, it is only necessary to use the SUPREM program, which considers the threshold voltage from a MOS capacitor point of view [5.3]. This method is valid because short channel effects such as drain-induced barrier lowering [5.4] are absent in this case. It would be a waste of time to use two-dimensional numerical programs in this case. There may be cases in which only rough estimates are needed to evaluate whether a particular idea

is feasible. In such a case, the simplest procedures that satisfy the relaxed accuracy requirements should be used. When the idea is considered feasible, and more detailed study is desired, then more sophisticated simulation tools will be used. Also, as mentioned in discussion 1), the simplest procedure usually provides the most basic and easily understood physical insight of the problem. In general, the idea is to choose the tool that provides the necessary accuracy, but not more.

5.2 Outline of the case studies

The following chapters are case studies in which simulations are used in VLSI device development. In Chapter 6, we present the application of SUPREM III in process development. The accuracy of the program is improved by adjusting the model parameters to fit experimental data. In Chapter 7, we discuss basic device physics for process engineers, and how to generate device parameters by using simulations. The relationship between device characteristics and process parameters are presented. Chapter 8 studies the drain-induced barrier lowering effect in shot channel MOSFET's. In Chapter 9, a combination of simulations and experiments are used to study in detail the lightly-doped drain (LDD) structure. The physics of the structure are revealed through simulations. Chapter 10 presents the use of CAD tools in analyzing a special structure, which is the trench isolation in CMOS. This chapter shows how simulations have identified a potential problem, and provided recommendations on how to minimize such problem. Chapter 11 shows the use of simulation tools in the development of new isolation structures such as SWAMI. Simulations are used to evaluate the performance of isolation structures. Chapter 12 presents how CAD is used in submicron CMOS transistor design. The issues of subthreshold leakage and drain-induced barrier lowering, as well as other concerns in short channel MOSFET's are discussed. Chapter 13 presents a methodology to systematically optimize a p-channel MOSFET with n^- pockets, using PISCES simulations. Chapter 14 shows how simulations can be used to optimize the scaling of a process to achieve higher circuit performance and minimize process complexity. Chapter 15 presents the simulation of parasitics in circuits. Parasitic resistances and capacitances are simulated in 2-D and 3-D, which provide very important information essential for circuit simulations. The physics of parasitic capacitances is also discussed.

References

[5.1] K. M. Cham and S. Y. Chiang, "Device Design for the Submicrometer P-Channel FET with n^+ Polysilicon Gate," *IEEE Trans. on Electron Devices*, **ED-31**, July 1984, pp. 964-968.

[5.2] G. E. P. Box, W. G. Hunter, and J. S. Hunter, *Statistics for Experimenters*, NY:John Wiley & Sons, 1978.

[5.3] S. M. Sze, *Physics of Semiconductor Devices*, 2nd ed., NY:Wiley Interscience, 1981.

[5.4] R. R. Troutman, "VLSI Limitations from Drain-Induced Barrier Lowering," *Trans. on Electron Devices*, **ED-27**, April 1979, pp. 461-468.

Chapter 6
SUPREM III Application

6.1 Introduction

SUPREM III has emerged as the most widely used process simulator. The authors of SUPREM III have attempted to include the most up-to-date physically based process models that are currently available and suitable for computer simulation. Innumerable research hours have gone into the development of the various models and fitting parameters. However a silicon process, whether it is bipolar, NMOS, or CMOS, is an extremely complicated entity. There are several problems inherent in the SUPREM process models.

1. All models are based on a limited data set. It would be impossible to accumulate data for the countless different combinations of process conditions that are currently used or may be used in the future. One should be especially careful when using SUPREM to make predictions for process steps that lie outside the range of the measured data used to formulate the process models. This has implications for shallow junctions, thin oxides and low temperature diffusion. All of which are important in advanced processes.

2. The data used to develop the process models is usually derived from carefully controlled experiments involving a small number of process steps.

However, in a real process, the many different process steps may interact with each other and produce unexpected results.

3. The models used in SUPREM III are physically based. Therefore interpolation between data points, where the same physics applies, should be acceptable. However it is possible that the process models are not based on complete or correct physical understanding. In this case, extrapolation between data points may not be accurate and slightly different process steps may produce unexpected results.

4. SUPREM III is inherently one-dimensional. However, as process dimensions shrink, two-dimensional effects become important. Therefore the dopant distribution in one area of the device structure becomes a function of the conditions present in adjacent regions (see Fig. 2.1). This is especially true for oxidation and diffusion which affect the point defect concentrations in adjacent semiconductor regions. The point defect concentrations then affect the diffusion processes in the adjacent regions.

For all the above reasons, the program user should have a healthy scepticism of the predictions from SUPREM. One should put confidence in the results only after they have been bench-marked against measurements, both electrical and analytical.

Typically, one is interested in predicting a particular set of electrical parameters for a particular process or class of processes. This set of electrical parameters may be very sensitive to some subset of the process steps. For example, MOS device parameters are very sensitive to the channel doping profile, gate oxide thickness and source/drain profiles. Latch-up, in a CMOS process, is most sensitive to the n-well and/or p-well profiles, and the substrate or epi doping. Bipolar device performance is most sensitive to the base doping profile. In each of these cases, a different set of process models have the most influence on the predicted electrical parameters.

In this chapter, we present some examples of the limitations of the process models in SUPREM that have been encountered in modeling device characteristics for CMOS processes. Suggestions are made on how to work around these limitations. Similar examples could be found for other sets of electrical parameters.

6.2 Boron Implant Profiles

SUPREM III uses the Pearson IV profile type for all implant profiles unless a different function is specified. In general the Pearson IV profile can do a good job of fitting measured implant profiles. However, boron implants into single crystal silicon produce relatively complex profile shapes [6.1-6.5]. Most of the problem is due to channeling where the implanted ions follow open corridors through the crystal array. This produces a long "tail" on the implant. The situation is further complicated since the exact nature of the profile is sensitive to the implant angle and to the rotation of the wafer about the implant angle.

Fig. 6.1 shows a measured boron implant profile for 30 KeV at 7° implant angle. The dashed line in this figure is one attempt to fit a Pearson IV profile to the implant profile. The fit is fairly good for the upper two orders of

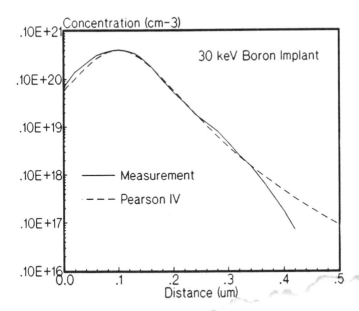

Fig. 6.1. Pearson IV Profile Fitted to a Measured Boron Implant Profile. The measured profile is from SIMS.

magnitude. However the measured and implanted profiles diverge at lower doping levels. Therefore this particular Pearson IV profile would work well for a channel implant where only the top two orders of magnitude are important. However, it would be very poor for a source/drain implant where the top five orders of magnitude are important. The Pearson IV profile would predict an excessive junction depth. Therefore one must be very careful using published Pearson IV coefficients since the fit may not be appropriate for all applications.

If one wished to represent the doping profile in Fig. 6.1 over a wider range of doping, a exponential tail on the profile would fit better than the Pearson IV. However, the Pearson IV profile with exponential tail is not an option supported by SUPREM III. Fig. 2.2 shows a comparison between the Pearson IV and Pearson IV with exponential tail.

6.3 Thin Oxide Growth

The growth of thin oxides is one of the most difficult aspects of IC processing to model [6.6-6.9]. The growth of the first 7 nm of oxide is very dependent on pre-oxidation cleans, the ambient present as the wafers are pushed into the furnace, pre-heat cycles, etc. Fig. 6.2 shows some data taken at Hewlett-Packard Laboratories for thin oxide growth at temperatures from 850 °C to 1000 °C in dry oxygen ambient. At least two effects are displayed in this data.

1. There is an apparent initial oxide thickness that is present even for "zero" time oxidation cycles. The log/log plot in Fig. 6.2a shows this effect. This initial oxide layer may result from the pre-oxidation clean and oxide growth during the wafer push.

2. An enhanced growth rate is present for thin oxides. This is easily observed when the data is presented on a linear/linear scale as in Fig. 6.2b. The slope of the thickness versus time plot is larger for thin oxides.

SUPREM III models the enhanced growth rate of thin oxides with the the following equation.

SUPREM III Application

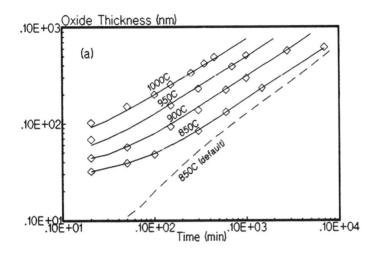

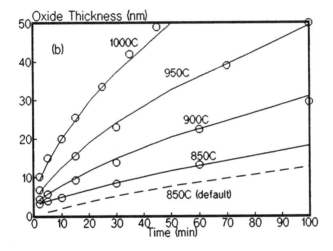

Fig. 6.2. SUPREM Oxidation Model Fitted to Measurements. The data points are measurements of dry oxidations performed at Hewlett Packard Laboratories. The solid lines are from SUPREM III using optimized coefficients for the thin oxide growth model. The dashed line shows the result from SUPREM III using the default thin oxide coefficients. Curve (a) is a log-log plot. Curve (b) is the same data on a linear-linear plot.

$$\frac{dx}{dt} = \frac{A}{2X_0 + B} + K \exp\left(-\frac{X_0}{L}\right) \qquad (6.1)$$

Here dx/dt is the oxide growth rate. A, A/B and K are all singly activated functions of temperature. The prediction of the default parameters for 850 °C is shown in Fig. 6.2. Obviously the parameters in this model need some modification to play-back the measurements. Also the initial oxide thicknesses must be introduced artificially. The solid lines in Fig. 6.2 result form the following set of parameters.

$$\frac{B}{A} = 6.18 \times 10^7 \text{ nm/min} \exp\left(-2.0 \frac{eV}{kT}\right) \qquad (6.2)$$

$$K = 5.2 \times 10^7 \text{ nm/min} \exp\left(-1.88 \frac{eV}{kT}\right) \qquad (6.3)$$

$$L = 13.8 \text{ nm} \qquad (6.4)$$

Initial Oxide Thickness

2.8 nm 850 °C
3.4 nm 900 °C
4.0 nm 950 °C
5.5 nm 1000 °C

The fit to the measurements is quite good. Here the linear rate coefficients are the same as the default values in SUPREM. However the value of L the parameters in K are significantly different from the default values.

For these calculations, the initial oxide thickness (Fig. 6.2) was deposited prior to the oxidation. This is not always desirable. An example is the gate oxidation in a typical CMOS process. This oxidation is usually done immediately after the shallow channel boron (or blanket) implant. This implant adjusts the threshold for both the n and p-channel devices. During the gate oxidation it is important to correctly model the incorporation of boron into the gate oxide (see section 6.4). Thus the entire oxide should be grown rather than depositing the first few nanometers. For this case the initial oxidation cycle can be lengthened to provide the initial oxide. For example, the coefficients listed above yield 2.8 nm in 12 minutes at 850 °C. Thus any 850 °C dry oxidation on

SUPREM III Application

bare silicon can be lengthened by 12 minutes to provide the initial oxide thickness. This slight additional thermal cycle will have very little effect on the resulting boron profile.

6.4 Oxygen Enhanced Diffusion of Boron

For low temperature oxidations, the diffusion constant for boron is dominated by the contribution from oxygen enhanced diffusion (OED) [6.11-6.20]. This is quite important for advanced CMOS processes. The gate oxide is usually grown after the shallow boron channel (or blanket) implant. This implant must be shallow to control leakage in the p-channel device. A significant fraction of the channel implant is "sucked up" into the gate oxide as it is grown. This boron is inactive in the gate oxide and does not contribute to the threshold adjustment. It is very important that the simulation predict the correct amount of boron remaining in the silicon after the gate oxidation or else the predicted threshold voltages will be incorrect.

The flow of boron into the gate oxide is controlled primarily by two parameters: the segregation constant (m) for the silicon/oxide interface and the diffusion constant (D) for boron during the gate oxidation (see sec. 2.2). The segregation constant and diffusion constant are determined by the following equations and parameters. Here the segregation constant is the ratio of the equilibrium dopant concentration in the oxide to that in the silicon.

$$m = m_0 \exp\left(\frac{E_m}{kT}\right) \tag{6.5}$$

$$D = D_N + D_{ox} \tag{6.6}$$

$$D_{ox} = D_I f_{ii} K \exp\left(-\frac{x}{L}\right)\left(\frac{dx}{dt}\right)^{0.5} f_{HCl} \tag{6.7}$$

$$f_{ii} = f_{ii0} \exp\left(-\frac{f_{iiE}}{kT}\right) \tag{6.8}$$

An experimental n-well CMOS process was developed at Hewlett Packard Laboratories using a 25 nm gate oxide grown at 850 °C. At this temperature

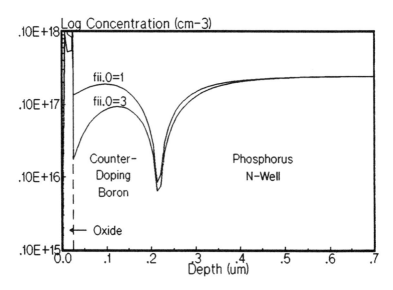

Fig. 6.3. SUPREM III Simulation of Channel Doping Profile of P-channel Transistor. This graph compares the predicted channel profile for two values of the parameter f_{ii0}. The segregation constant was 10 for each simulation.

the boron diffusion constant is dominated by the contribution from OED. Fig. 6.3 shows the simulated channel doping profile in the p-channel device for two values of the parameter f_{ii0} which is a dimensionless parameter defined differently for each dopant species. The parameter f_{iiE} is kept at its default value (0.57 eV). It is clear from Fig. 6.3 that the predicted channel doping profile is quite sensitive to the value of the OED parameters.

Fig. 6.4 shows the fraction of the channel boron implant that migrates to the gate oxide as a function of f_{ii}. One curve in Fig 6.4 uses the default segregation constant m which is about 10 at 850 °C. The other curve fixes m at a value of 3.

Fig. 6.5 shows how the long-channel extrapolated threshold voltage for the n and p-channel device varies with f_{ii0}. Here the threshold voltages were calculated by three techniques. One of these used the electrical parameter extraction features of the SUPREM III program. In this mode, SUPREM solves the one-dimensional Poisson equation for a given gate and substrate

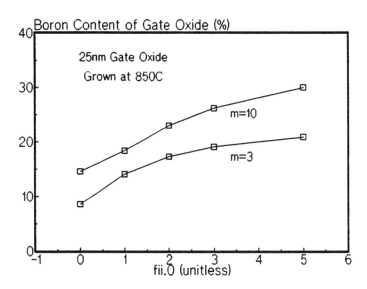

Fig. 6.4. Boron Content of Gate Oxide. This plot shows the fraction of the shallow channel boron implant that is incorporated into the gate oxide as a function of the parameter f_{ii0} for two values of the segregation constant m.

bias. From this, the channel charge is calculated as function of gate bias. Plotting channel charge versus gate bias allows one to determine the extrapolated threshold voltage.

The other two curves in Fig. 6.5 were obtained from the device simulation program PISCES IIB. PISCES IIB is used to calculate the drain current for each gate bias. The extrapolated threshold is then obtained in the usual way. In one curve, a constant carrier mobility was used. In the other, a field dependent mobility was specified (see Section 3.3). The constant mobility threshold voltages agree well with the threshold voltages from SUPREM. The field dependent mobility changes the shape of the drain current versus gate bias curve. This shifts the extrapolated threshold to a slightly lower voltage. It is these values that should be compared to measurements which are also influenced by the field dependence of the carrier mobility.

It is obvious from Fig. 6.5 that the values chosen for the OED parameters and the segregation constant have a significant effect on the predicted

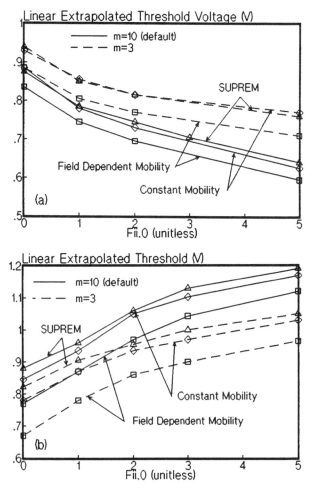

Fig. 6.5. Simulated Threshold Voltage. The SUPREM channel doping profiles are used to simulate the long channel threshold voltage. In all cases the threshold voltage was extrapolated from the linear region of the I_D versus V_G curve at low drain bias. The threshold was calculated in three ways: (i) SUPREM III which solves the 1-D Poisson equation. (ii) PISCES IIB using a constant carrier mobility. (iii) PISCES IIB using a field dependent carrier mobility. The predicted threshold voltages are presented as a function of the parameter f_{ii0} for two values of the segregation constant. (a) N-channel Device (b) P-Channel Device.

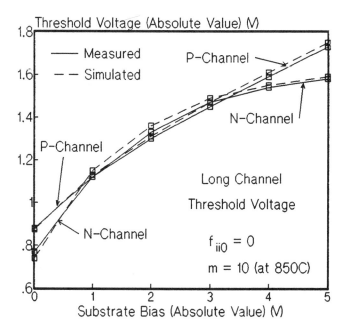

Fig. 6.6. Long Channel Threshold Voltage versus Substrate Bias. The simulated threshold voltages were calculated with SUPREM III doping profiles in the PISCES IIB device simulator using the field dependent mobility model. For both the measured and simulated cases, the threshold voltage was extrapolated from the linear region of the I_D versus V_G curve at low drain bias.

threshold voltages. The measured nominal extrapolated threshold voltages for this process are 0.77 V for the n-channel device and -0.88 V for the p-channel. Only the threshold voltages simulated using the field dependent mobility model in PISCES should be compared directly to the measured values. For the range of segregation constant between 3 and 10 in Fig. 6.5, there is a range of f_{ii0} that would predict the correct threshold voltages. This range is approximately $1 \leq f_{ii0} \leq 2$ which is significantly different from the default value of 4.1. A reasonably good choice is to use $m = 10$ and $f_{ii0} = 1$. The resulting profile predicts the correct dependence of threshold on substrate bias as shown in Fig. 6.6. This indicates that the simulated channel profile is probably a good representation of the actual profile.

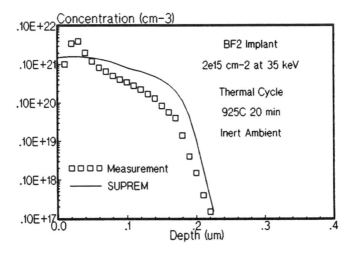

Fig. 6.7. Comparison of SUPREM Simulation and Measurements for a P+ Profile formed by BF_2 Implantation and Anneal. The measurements were made by SIMS.

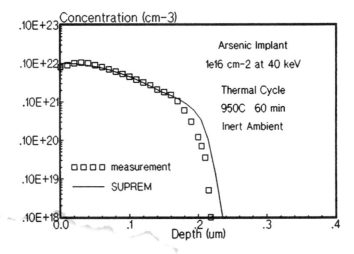

Fig. 6.8. Comparison of SUPREM Simulation and Measurements for a N+ Profile formed by Arsenic Implantation and Anneal. The measurements were made by SIMS.

6.5 Shallow Junctions

In advanced MOS processes, the source/drain junction depth becomes an important parameter. Short channel effects are sensitive to this parameter. Shallow junctions are usually made by implanting arsenic (n^+) [6.22] or BF_2 (p^+) [6.3] and annealing at relatively low temperatures (850 °C to 950 °C) in inert or dry ambient. For inert ambient anneals, the n^+ or p^+ profile is determined predominantly by concentration enhanced diffusion (CED) effects. This is modeled in SUPREM III by letting the diffusion constant be a function of the local free carrier density (see Eqs. (2.20) and (2.21)) which depends on the doping density. This causes a high concentration arsenic or boron profile to diffuse more quickly than a low concentration profile.

Figs. 6.7 and 6.8 compare SUPREM III profiles with measured profiles for some typical source/drain implants and anneals. In general the agreement is quite good. This is usually true for arsenic and boron (BF_2 implanted) profiles. Thus the models and parameters for CED of boron and arsenic appear to be reasonably accurate.

The situation is not as good for phosphorus. The diffusion of phosphorus in heavily doped silicon has always posed problems for simulation and our understanding is still incomplete. This arises in advanced MOS processes for "double diffused" junctions. Here a high dose of arsenic and a lower dose of phosphorus are implanted to form the n^+ drain. This produces a more graded junction than arsenic alone, thus reducing the electric field at high drain bias (see Chapter 12). Our experience indicates that the models in SUPREM for phosphorus diffusion are not reliable for predicting the correct profiles for shallow double diffused junctions. Usually the simulated profile is shallower than the actual profile.

References

[6.1] H. Ryssel, et al, "Range Parameters of Boron Implanted into Silicon," *Appl. Phys.* 24, 1981, pp. 39-43.

[6.2] M. Simard-Normandin and C. Slaby, "Empirical Modeling of Low Energy Boron Implants into Silicon," *J. Electrochem. Soc.* 132(9). 1985, pp. 2218-2223.

[6.3] R.G. Wilson, "Boron, Fluorine, and Carrier Profiles for B and BF_2 Implants into Crystalline and Amorphous Si," *J. Appl. Phys.* 54(12), 1983, pp. 6879-6889.

[6.4] A.E. Michel, et al, "Channeling in Low Energy Boron Ion Implantation," *Appl. Phys. Lett.* 44(4), 1984, pp. 404-406.

[6.5] F. Jahnel, et al, "Description of Arsenic and Boron Profiles Implanted in SiO_2, Si_3N_4 and Si Using Pearson Distributions with Four Moments," *Nucl. Instrum. and Methods* (Netherlands) 182/183, 1981, pp. 223-229.

[6.6] Y.J. Van der Meulen, "Kinetics of Thermal Growth of Ultra-Thin Layers of SiO_2 on Silicon, Part 1. Experiment," *J. Electrochem. Soc.*, 119, 1972, p. 530.

[6.7] R. Ghez and Y.J. Van der Meulen, "Kinetics of Thermal Growth of Ultra-Thin Layers of SiO_2 on Silicon, Part 2. Theory," *J. Electrochem. Soc.*, 119, 1972, p. 1100.

[6.8] E.A. Irene, "Silicon Oxidation Studies: Some Aspects of the Initial Oxide Regime," *Appl. Phys. Lett.* 33, 1978, p. 424.

[6.9] G.F. Derbenwick and R.E. Anderson, "Rapid Initial Thermal Oxidation of Silicon," *J. Electrochem. Soc.*, 1983.

[6.10] H.Z. Massoud, "Thermal Oxidation of Silicon in Dry Oxygen - Growth and Charge Characterization in the Thin Regime," Stanford Electronics Laboratories, TR G502-1, Stanford University, Stanford Calif., 1983.

[6.11] D.A. Antoniadis, A.M. Lin, and R.W. Dutton, "Oxidation-Enhanced Diffusion of Arsenic and Phosphorus in Near-Intrinsic (100) Silicon," *Appl. Phys. Lett.* 33, 1978, p. 1030.

[6.12] K. Taniguchi, K. Kurosawa, and M. Kashiwagi, "Oxidation Enhanced Diffusion of Boron and Phosphorus in (100) Silicon," *J. Electrochem.*

Soc. 127, 1980, p. 2243.

[6.13] A.M. Lin, D.A. Antoniadis, and R.W. Dutton, "The Oxidation Rate Dependence of Oxidation-Enhanced Diffusion of Boron and Phosphorus in Silicon," *J. Electrochem. Soc.* 128, 1981, p. 1131.

[6.14] Y. Ishikawa, et al, "The Enhanced Diffusion of Arsenic and Phosphorus in Silicon by Thermal Oxidation," *J. Electrochem. Soc.*

[6.15] D.A. Antoniadis, A.G. Gonzalez, and R.W. Dutton, "Boron in Near-Intrinsic (100) and (111) Silicon under Inert and Oxidizing Ambients - Diffusion and Segregation," *J. Electrochem. Soc.* 125, 1978, p. 813.

[6.16] J.W. Colby and L.E. Katz, "Boron Segregation at Si-SiO$_2$ Interface as a Function of Temperature and Orientation," *J. Electrochem. Soc.* 123, 1976, pp. 409-412.

[6.17] R.B. Fair and J.C. Tsai, "Theory and Direct Measurement of Boron Segregation in SiO$_2$ During Dry, Near Dry, and Wet O$_2$ Oxidation," *J. Electrochem. Soc.* 125, 1978, pp. 2050-2058.

[6.18] A.S. Grove, O. Leistiko, and C.T. Sah, "Redistribution of Acceptor and Donor Impurities During Thermal Oxidation of Silicon," *J. Appl. Phys.* 35, 1964, pp. 2695-2701.

[6.19] B.E. Deal, A.S. Grove, E.H. Snow, and C.T. Sah, "Observation of Impurity Redistribution During Thermal Oxidation of Silicon Using the MOS Structure," *J. Electrochem. Soc.* 112, 1981, pp. 1101-1106.

[6.20] Y. Ishikawa, et al, "The Enhanced Diffusion of Arsenic and Phosphorus in Silicon by Thermal Oxidation," *J. Electrochem. Soc.* 129, 1982, p. 645.

[6.21] R.B. Fair, "Oxidation, Impurity Diffusion, and Defect Growth in Silicon - An Overview," *J. Electrochem. Soc.* 128, 1981, p. 1360.

[6.22] M.Y. Tsai, F.F. Morehead, J.E. Baglin, and A.E. Michel, "Shallow Junctions by High-Dose As Implants in Si: Experiments and Modeling," *J. Appl. Phys.* 51(6), 1980, pp.3230-3235.

Chapter 7
Simulation Techniques for Advanced Device Development

In this chapter, the basic simulation techniques for advanced MOS device development will be described. First of all, the basic device physics of MOSFET is presented. The discussion will be in very simple terms, although sufficient to allow the process engineers to understand the basic characteristics of MOSFET's and their significance. The techniques of generating the device parameters are then presented. Also to be discussed are the short channel effects such as drain-induced barrier lowering. Simulations are used to reveal details of these phenomena. The relationship between process parameters and device characteristics are discussed. Simulated results are compared with experimental results. The SUPREM, GEMINI and PISCES programs are used for simulating the device characteristics.

7.1 Device Physics for Process Development

The physics of the MOSFET has been discussed in detail in numerous books and papers [7.1]-[7.5]. The purpose of this section is not to reproduce those discussions, but rather to highlight in simple terms the critical

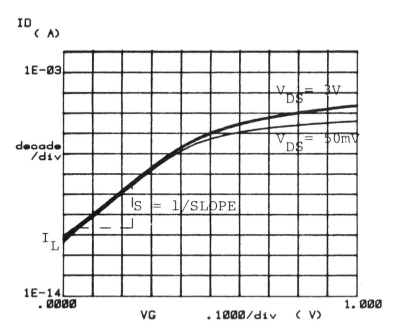

Fig. 7.1. Drain current (log scale) vs. gate bias for a long channel MOSFET, with drain to source bias (V_{DS}) of 50 mV and 3 V.

parameters for process development, and discuss briefly their significance to the circuit performance.

The electrical characteristics of MOSFET's can be separated into two regions. One region is the subthreshold region, where the drain current of the transistor is exponentially dependent on the gate voltage. Fig. 7.1 shows the measured subthreshold characteristics of a long channel MOSFET. In this region, the most interesting parameters are the subthreshold slope, threshold voltage, and the residual current at zero gate bias.

The subthreshold slope (S) is conventionally expressed as the inverse of the slope of the drain to source current (in logarithmic scale) versus the gate to source bias. S measures how fast the drain current can be turned off as a function of gate bias. It is typically between 80 to 120 mV/decade. The steeper the slope (hence smaller S), the less gate voltage sweep will be necessary to turn the current off to a desired level. In this case, the threshold voltage

Simulation Techniques

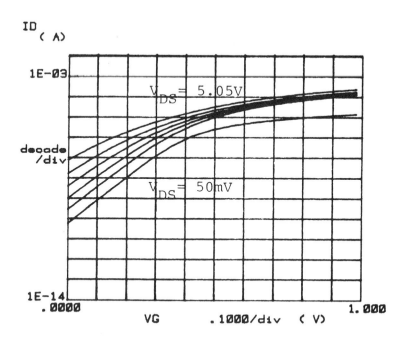

Fig. 7.2. Drain current (log. scale) vs. gate bias for a short channel MOSFET with V_{DS} increasing from 50 mV to 5.05 V.

of the device can be reduced, meaning a larger current drive for the same circuit bias voltage, and thereby providing higher performance when the gate and drain are in the active range. The subthreshold slope also determines the residual drain current (I_L) for the same threshold voltage, as shown in the figure. A steeper slope reduces the residual current, if the residual current is mainly due to diffusion of the carriers through the potential barrier between the source and the drain and not due to other factors such as leaky junctions. The subthreshold slope degrades as the transistor channel length is reduced to the extent that short channel effects are significant. Fig. 7.2 shows the subthreshold characteristics of a short channel transistor. The degradation of the slope is more serious for higher drain biases, resulting in significant residual leakage current at zero gate bias. This degradation is due to the drain-induced barrier lowering (DIBL) effect. The physics of the subthreshold characteristics have been discussed in detail in many articles [7.3]-[7.6]. It is useful to point out that in the simplest case where DIBL effects are absent, the subthreshold slope can be approximately expressed as

$$S = 2.3\frac{kT}{q}[1 + \frac{C_D}{C_{ox}}] \tag{7.1}$$

where C_D and C_{ox} are the capacitance of the depletion layer and the gate oxide capacitance respectively. It can be seen here that by reducing the gate oxide thickness, the subthreshold slope can be reduced significantly. The absolute minimum value of S is 59.6 mV/dec at room temperature. The extraction of the subthreshold characteristics for short channel devices using simulations will be presented in section 7.3.

For a given subthreshold slope, the noise margin and residual current are determined by the threshold voltage. (For short channel devices, the subthreshold slope and threshold voltage are related. The slope degrades, especially under a large drain bias, as the threshold voltage is reduced by using a lower impurity concentration in the channel region. This is due to the DIBL effect.) The compromise in targeting the threshold voltage is between the current drive, noise margin and residual leakage current. Here the noise margin is not a major concern since CMOS circuits are very tolerant in this respect. The residual current is a major concern in submicron CMOS processes. In the case where low power operation is desired, such as in a battery operation environment, the residual current can contribute greatly to the standby power. Thus it is desired that the residual current be minimized. This can be realized by choosing a threshold voltage high enough to prevent significant off current. A higher threshold voltage reduces the residual current not only because of the magnitude of the threshold voltage, which provides more turn-off margin for the transistor, but also because of the higher channel impurity concentrations which reduces the DIBL effects. Hence the subthreshold slope also improves.

In the "turned-on" region, the transistor characteristics can be described by the following simple equations for long channel lengths, typically larger than 3 μm, and drain bias below saturation [7.1]:

$$I_{DS} = \frac{W}{L}\mu C_{ox}(V_G - 2\psi_B - \frac{V_D}{2})V_D$$

$$- \frac{2W}{3L}\mu\sqrt{2\varepsilon q N_A}[(V_D + 2\psi_B)^{3/2} - (2\psi_B)^{3/2}] \quad (7.2)$$

$$I_{DS} \approx \frac{W}{L}\mu C_{ox}[(V_{GS} - V_T)V_{DS} - \frac{1}{2}V_{DS}^2] \quad (7.3)$$

for low doping concentrations and thin gate oxides;

$$I_{DSAT} = \frac{W}{2L}\mu C_{ox}(V_{GS} - V_T)^2 \quad (7.4)$$

at saturation. Here μ is the mobility of the carriers in the channel. Obviously it is necessary to have a good mobility model to accurately describe the device characteristics. The mobility is dependent on the vertical electric field in the channel, and for short channel MOSFET's, the effect of the lateral field on the mobility has to be taken into consideration [7.7] (see Ch. 3, Eq. (3.15)). In very short channel transistors, the major deviation from the above equation is the velocity saturation phenomenon, where I_{DS} is not proportional to $(V_{GS}-V_T)^2$ in the saturation region, but rather proportional to $(V_{GS}-V_T)$, as given by the following equation [7.8]:

$$I_{DSAT} = WC_{ox}V_s(V_{GS} - V_T) \quad (7.5)$$

where V_s is the saturation velocity of the carriers. Fig. 7.3 shows typical transistor I-V characteristics of a long channel MOSFET, and Fig. 7.4 for a short channel device.

It is useful to be able to predict the performance of the transistors, so that the circuit performance can be predicted before fabrication is actually done. There is always a compromise between transistor performance and process complexity. If the device performance can be predicted, then an optimized process and device design can be developed. There are many other device characteristics issues that the engineer has to be concerned about. One example would be the drain to source breakdown of the MOSFET due to the parasitic bipolar effect between the source and drain [7.9],[7.10]. Fig. 7.5 shows the breakdown behavior of an n-channel transistor for a channel length of one micron. Fig. 7.6 shows the source-drain breakdown voltage (BV_{DS}) versus the

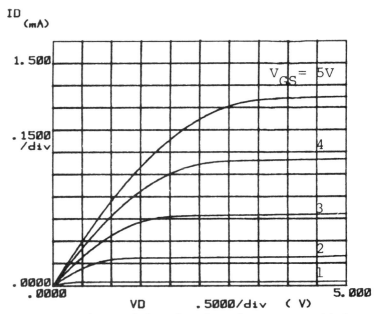

Fig. 7.3. *I-V* characteristics of n-channel MOSFET with long channel length (≈ 5 μm).

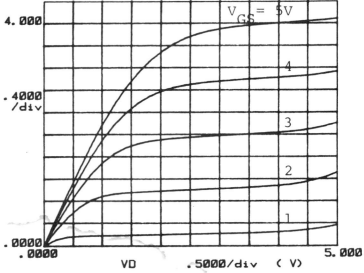

Fig. 7.4. *I-V* characteristics of n-channel MOSFET with short channel length (≈ 0.7 μm).

Simulation Techniques

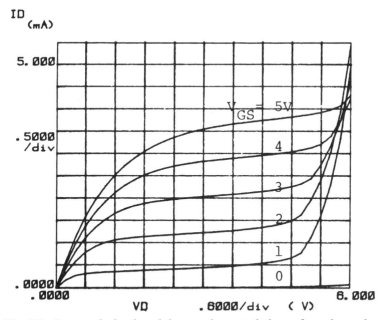

Fig. 7.5. Source-drain breakdown characteristics of n-channel MOSFET with short channel length.

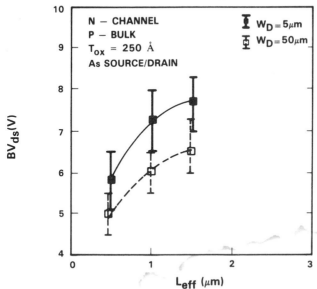

Fig. 7.6. Breakdown voltage (BV_{DS}) of n-channel MOSFET vs. effective channel length (L_{eff}) and width (W_D).

effective channel length for n-channel MOSFET's with conventional source/drain structure using arsenic implant. This causes serious reliability problems in the circuit. Many techniques such as using epitaxial silicon on low resistivity substrate, graded junctions, LDD structure [7.11], etc. have been implemented to minimize this problem. Since this is a complicated phenomenon, experimental data is often used as reference during process development. Simulations can provide physical insight into this problem.

ISSUES	FACTORS	COMPROMISES
PERFORMANCE (CURRENT DRIVE)	CHANNEL LENGTH GATE OXIDE THICK. ISOLATION CHANNEL IMPLANTS SERIES RESISTANCE	PUNCHTHROUGH PROCESS COMPLEXITY
PUNCHTHROUGH	CHANNEL LENGTH GATE OXIDE THICK. GATE MATERIAL CHANNEL IMPLANTS JUNCTION DEPTHS	PERFORMANCE PARASITIC CAPAC. SERIES RESISTANCE BODY EFFECT
HOT ELECTRON/ BREAKDOWN (N-CHANNEL)	S/D STRUCTURE SPACER WIDTH EPI-WAFERS GATE OXIDE THICK.	PROCESS COMPLEXITY SERIES RESISTANCE PERFORMANCE
DIFFUSION CAPAC. PERIPHERAL: AREAL:	FIELD IMPLANT N-CH. IMPLANT	FIELD LEAKAGE PUNCHTHROUGH

Table 7.1. Device optimization considerations.

The hot electron effect is another concern in VLSI. As the channel lengths are scaled down faster than the power supply voltage, together with scaled down oxide thickness and junction depths, the electric field at the drain is increased to the extent that hot electrons cause degradation of the device performance [7.12]-[7.14]. In this book, a detailed study of this effect in the LDD structure is presented. Another phenomenon of major concern is latch-up in CMOS. Simplified models [7.15] as well as sophisticated numerical programs [7.16],[7.17] have been used to simulate the latch-up characteristics.

Simulation Techniques 175

This is another example of CAD in device study. Table 7.1 shows the device parameters that have to be optimized (first column), the factors affecting the parameters (second column), and the compromises (third column) that have to be made. For example, reducing the channel length will increase the current drive, but may cause punchthrough problems. The table shows that device optimization is a complicated process in VLSI development.

7.2 CAD Tools for Simulation of Device Parameters

In this section, the simulation tools that are used at Hewlett-Packard Laboratory for the simulation of device parameters are discussed. From the previous discussion on the methodology of simulations in process development, it is clear that we need to choose the programs that will provide the necessary information with the desired accuracy in the shortest time. This means that we have to use different programs in different situations, even for generating the same parameter. The most significant differences between devices are long channel versus short channel. The following discussions will provide general guidelines in choosing the programs for device simulations. Note that this section is concerned only with transistor parameters. Chapter 15 will discuss the programs that are appropriate for generating the parasitic resistances and capacitances.

Table 7.2 shows the two categories of programs that are used for process simulations and device simulations. (The SEDAN program have been implemented at Hewlett-Packard Laboratories, but will not be discussed in this book.) In most cases, the simulated impurity profiles and insulator structures can be transferred to the device simulation programs for generating the device parameters. Table 7.3 summarizes the programs that are appropriate for generating the critical parameters for process development. The table is listed in order of increasing complexity, which means increasing computing time. This does not necessarily mean increasing accuracy since it depends on the particular device being considered. The SUPREM program can be used to generate threshold voltages of long channel MOSFET's, or any device that exhibits negligible short channel behavior. This is also the fastest

PROCESS	DEVICE
SUPREM II	GEMINI
SUPREM III	PISCES
SUPRA	CADDET
SOAP	SEDAN

Table 7.2. Process and device simulation programs used at Hewlett-Packard Laboratories.

PROGRAMS	DEVICE PARAM.	REMARKS
SUPREM II & III	VT	LONG CHANNEL, 1-D
GEMINI	VT, SUBVT, VPT	GAUSSIAN PROFILE
SUPREM II + GEMINI	VT, SUBVT, VPT	ARB. 1-D PROFILE
SUPRA + GEMINI	VT, SUBVT, VPT	ARB. 2-D PROFILE
SOAP + SUPRA +GEMINI	VT, SUBVT, VPT	ARB. 2-D PROFILE & ISOLATION
CADDET	VT, SUBVT, VPT, I-V	GAUSSIAN PROFILE
PISCES	VT, SUBVT, VPT, I-V, TRANS.	GAUSSIAN PROFILE
SUPREM III + PISCES	VT, SUBVT, VPT, I-V, TRANS.	ARB. 1-D PROFILE
SUPREM IV + PISCES	VT, SUBVT, VPT, I-V, TRANS.	ARB. 2-D PROFILE & ISOLATION

Table 7.3. Programs used for generating the device parameters. Here VT is threshold voltage, SUBVT is subthreshold characteristics, I-V is current-voltage characteristics above threshold, VPT is punchthrough voltage and TRANS is transient analysis.

Simulation Techniques

approach to provide a preliminary set of data even for short channels, since it is often possible to make rough corrections on the long channel threshold voltage to provide an estimate of the short channel threshold voltage.

The GEMINI program has the next higher level of complexity. It solves the two-dimensional (2-D) Poisson equation and generates device parameters such as the subthreshold slope. In the case of short channel devices, where the threshold is very sensitive to the channel length and drain bias, it is necessary to use this program to calculate the threshold voltage. In this case, two-dimensional effects are dominant, with the drain bias affecting the potential

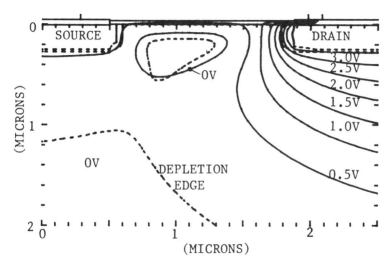

Fig. 7.7. Simulation of an n-channel MOSFET with channel length of 1.2 μm. The gate and drain bias are 0 and 3 V respectively.

barrier between the source and drain. Fig. 7.7 shows a simulation of a short channel transistor with a drain bias of 3 V. The depletion region at the drain extends into the channel area. This shows clearly that 2-D simulation is absolutely essential. The GEMINI program can provide a reasonable approximation of the channel, source and drain profiles, using Gaussian distributions.

The third degree of complexity is the combination of SUPREM II and GEMINI, which provides accurate vertical impurity profiles for the channel and the source/drain, during the device simulation. In this case, the channel and source/drain profiles are more realistic, since they are typically not Gaussian. For example, in short channel devices, a combination of shallow and deep implants is used to set the threshold voltage as well as the punchthrough voltage. Also, the source and drain profiles may be much steeper than a Gaussian profile, due to the thermal diffusion properties of the impurities, which depends on the impurity concentrations [7.18]. Fig. 7.8 and Fig. 7.9 show typical channel and source/drain profiles respectively, as simulated by SUPREM II. The source/drain impurity profile may not affect the threshold voltage simulation significantly, but will affect the DIBL and hot electron effect simulations.

The fourth degree of complexity is the combination of SUPRA and GEMINI, which provides not only the vertical impurity profile, but also the horizontal distribution at the source/drain. This is essential if one wants to simulate source/drain structures that are more complicated, such as the LDD [7.11] structure. SUPRA also allows one to simulate isolation structures, with certain limitations, and couple that structure to GEMINI for device simulations. This is useful for the simulation of narrow width effects or for the simulation of field parasitic transistors with advanced field isolation structures. For more complicated isolation structures such as SWAMI [7.19], it is necessary to couple the SOAP, SUPRA, and GEMINI programs to simulate the narrow width effects.

The CADDET program is used when it is necessary to investigate the full I-V characteristics of the device. The program does not accept SUPREM or SUPRA output files. Hence it is not as accurate as the SUPREM-GEMINI or SUPRA-GEMINI combinations in the simulation of the threshold voltage, unless the profiles are carefully specified by first running the SUPREM program and then approximating the profile with a Gaussian distribution. The applications of this program are described in Chapter 3, 9 and 14. The PISCES program is a two-dimensional, two carrier semiconductor device modeling program, which can simulate device characteristics under steady state or transient conditions. The program, when coupled with SUPREM III, provides much larger flexibility in the device structure as well as solution accuracy compared with the GEMINI and CADDET programs. The program is useful in simulating both the subthreshold and I-V characteristics of

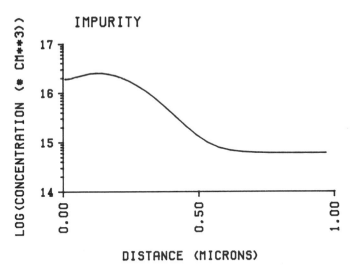

Fig. 7.8. Channel impurity profile of n-channel MOSFET, with a B_{11} implant of 5E11 cm^{-2} at 30 KeV and 4E11 cm^{-2} at 70 KeV.

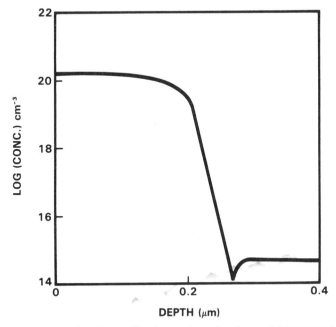

Fig. 7.9. Source/drain profile (Arsenic) of n-channel MOSFET.

transistors, as well as transient behaviors such as latchup. The disadvantage of PISCES is longer CPU time. The issue of CPU time is becoming less important as computer hardware improves continually. Also, the input file for PISCES can be simplified by using macro commands. Examples of PISCES applications will be presented in Chapters 3, 7, 8, 12, and 13. The coupling of PISCES and a 2-D process simulator such as SUPREM IV (being developed at Stanford University) offers the potential for the highest degree of accuracy and flexibility in the device simulation, provided that the individual programs have been verified.

7.3 Methods Of Generating Basic Device Parameters

In this section, methods of generating basic device parameters will be presented. Threshold voltage, subthreshold slope and punchthrough voltage will be simulated using different techniques. The assumptions used by the different programs will be discussed, so that the users are aware of the approximations being applied.

The first example is the simulation of the threshold voltage using the SUPREM program. In this case, the threshold voltage will correspond to the MOS capacitor threshold, which is very close to the long channel MOSFET threshold measured by the current versus gate voltage method. The model used is based on the full depletion approximation and the assumption of quasi-neutral impurity profiles [7.20]. The user inputs the process steps for the MOS structure. The program then calculates the impurity profile, and the threshold voltage as a function of the substrate bias. Fig. 7.10 shows the curve of simulated threshold voltages for p-channel transistors as a function of the counter-doping implant dose, together with experimental data. Good agreement is observed. This method of simulating the threshold voltage has the advantage of being the most simple, and the least time consuming. The limitation is that the threshold corresponds to the long channel threshold. This program is useful when one wants to look at the dependence of threshold voltage on the impurity profile without the complication of the short channel effects.

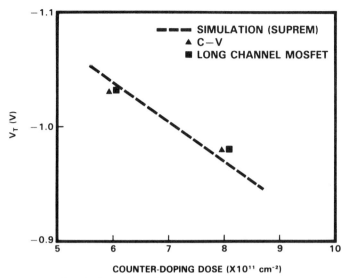

Fig. 7.10. Simulation of p-channel threshold voltage vs. counter-doping dose.

If the threshold voltage of a short channel transistor is to be simulated, then two-dimensional numerical analysis is necessary to provide good accuracy. The combination of SUPREM and GEMINI in most cases provides sufficient accuracy. The channel and source/drain implant profiles are first simulated by the SUPREM program. The structure of the transistor is defined by the GEMINI program. The impurity profiles at the channel and source/drain area are provided by the SUPREM data file after the SUPREM simulation is completed. Fig. 7.11 shows the bird's-eye-view of the impurity profile generated by the GEMINI program using the SUPREM data. The source/drain and channel profiles are displayed very clearly. The GEMINI program then solves Poisson's equation based on those profiles and bias conditions. It is important to know the threshold voltage of a short channel device biased at the power supply voltage V_{DD}, since this is the condition at which the device is biased when it is in the off state in an inverter circuit. For large drain bias, short channel effects are very significant and device characteristics have to be determined by 2-D simulations. The GEMINI program simulates the subthreshold characteristics of the transistor under this bias condition by

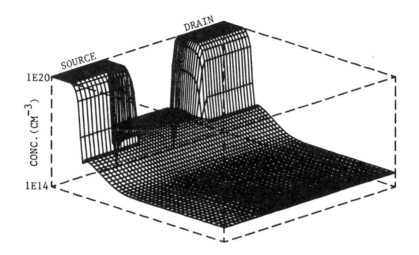

Fig. 7.11. Bird's-eye-view of the impurity profile (log. scale) for n-channel MOSFET. The substrate concentration is 6E14 cm^{-3}.

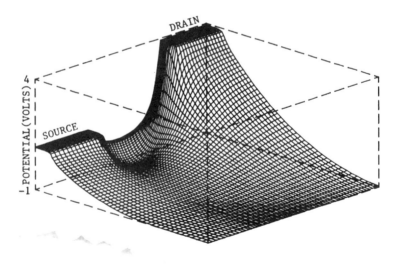

Fig. 7.12. Bird's-eye-view of the potential profile for n-channel MOSFET, with gate, drain and substrate bias of 0, 3, and -1 V respectively.

calculating the drain to source current versus gate bias, from which the subthreshold slope and threshold voltage can be extracted.

The GEMINI program solves Poisson's equation within the device structure and generates the potential profile. Fig. 7.12 shows the bird's-eye-view of the potential profile within the device structure. The effect of the drain bias in lowering the potential barrier can be observed qualitatively. In the GEMINI program, the quasi-Fermi level at the channel region is set equal to that of the drain bias. Hence under a large drain bias, the program underestimates the inversion charge after the device is turned on, and overestimates the band bending. The subthreshold current will continue to increase exponentially even after the threshold voltage is reached. Fig. 7.13 shows the simulation of the subthreshold characteristics of a n-channel MOSFET, together with the experimental data. The simulated data begins to deviate from the experimental data at high current levels, hence precautions need to be taken when interpreting the simulation results. It is useful to define a threshold current, from which the threshold voltage can be extracted. The conventional value of the threshold current is given by [7.21]

$$I_{TH} = (\frac{W_{eff}}{L_{eff}}) \times 10^{-7} \text{ A} \tag{7.6}$$

where L_{eff} and W_{eff} are the effective channel length and channel width respectively. If this value of threshold current is used, then the simulated threshold voltage will be too low, due to the deviation from experimental data at turn-on. This deviation is reduced for longer channel lengths. For example, at $L_{eff} = 2.5$ μm, the correction in the simulated threshold voltage is 0.03 V. In order to be consistent with experiment, it would be necessary to include this correction factor for short channel transistors. Another method would be to use a lower value for the threshold current during measurement. Taking the correction into account, the agreement between experiment and simulation is very good over a wide range of implant doses and energies (see Fig. 7.16 and 7.17 in section 4). The subthreshold slope can also be calculated from this simulation by taking the inverse slope of the log(I_{DS}) versus V_{GS} curve.

A more accurate method of simulating the subthreshold slope and threshold voltage of short channel devices is to use the PISCES and SUPREM III programs. The solution is valid for the entire range of bias conditions applicable to the device. Fig. 7.14 shows the result of a subthreshold simulation. The

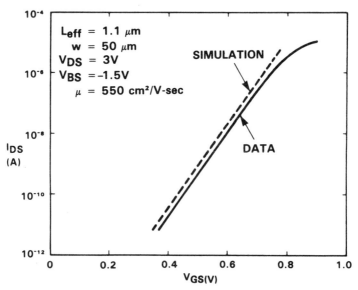

Fig. 7.13. MOSFET subthreshold characteristics simulation.

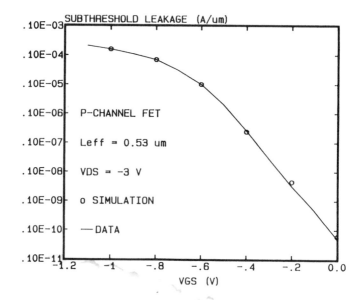

Fig. 7.14. Simulation of the subthreshold characteristics for a p-channel MOSFET, using the PISCES program.

Simulation Techniques   185

agreement is good from subthreshold to beyond threshold. The compromise is longer CPU time.

Although the SUPREM II-GEMINI combination is sufficient for many applications, sometimes it is necessary to simulate characteristics of devices with complicated source/drain or isolation structures. Examples are subthreshold leakage of p-channel transistor with n⁻ pocket [7.22] (see Chapter 13) and narrow width effects of LOCOS and SWAMI isolations (see Chapter 11). In these cases, the SUPREM III-PISCES or the SUPRA-GEMINI combination are needed.

In the design of short channel transistors, a major concern is the problem of "punchthrough", or more accurately, drain-induced barrier lowering (DIBL) [7.4]. The bias applied at the drain of the transistor has the effect of lowering the potential barrier between the source and the drain. "Punchthrough" traditionally means that the drain and source depletion regions are merged, and the maximum potential barrier to carriers is less than the junction "built-in" voltage. In this case, the bias at the drain can affect the potential distribution at the source. But the leakage current may still be negligible until the drain bias has caused significant lowering of the potential barrier height between the source and the channel. Since the diffusion current between the source and the drain is exponentially proportional to this barrier, the leakage current is very sensitive to the drain bias at short channels. The detailed physics of DIBL will be discussed in Chapter 8. Since leakage current is a major concern for low power operation, the DIBL effect is always simulated during process development. (This is especially important for VLSI device development where channel length is one micron or less.) This can be accomplished by using a combination of process simulator and device simulator. The process simulator generates the channel and source/drain profiles, and the device simulator simulates the DIBL effect as a function of the drain bias. Fig. 7.15 shows the simulated data of leakage current at zero gate bias vs. the drain bias, for an n-channel transistor with channel length of $0.6\,\mu$m using the SUPREM II and GEMINI programs. It is useful to define a "punchthrough voltage", which can be defined as the drain bias at which a certain level of drain current (punchthrough current) is observed at zero gate bias. The definition of this current level would depend on the application of the devices. Fig. 7.15

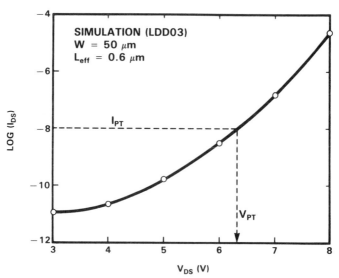

Fig. 7.15. N-channel MOSFET punchthrough characteristics.

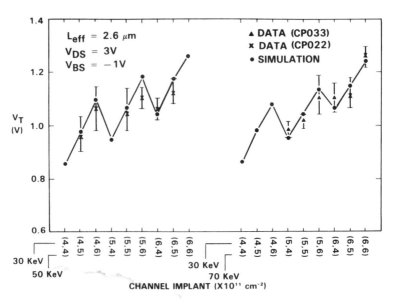

Fig. 7.16. Threshold voltage vs. channel implant for n-channel MOSFET with channel length of 2.6 μm.

Drain-Induced Barrier Lowering

Fig. 8.6, together with the potential contours. The punchthrough current will flow along the path which contains the minimum potential barrier height. Conventionally, the term "saddle point" is referred to the point along the bulk current path where the potential is minimum or maximum for electron or hole respectively. (The "saddle" shape can be visualized by combining Figures 8.4, 8.5 and 8.6.)

In this particular case, according to the simulation results, the punchthrough path occurs at the surface, and the minimum electron potential occurs at $x = 1.24 \mu m$ (point A in Fig. 8.6) for a punchthrough current of 1 nA and device width of 50 μm. From here on, the term "punchthrough point" refers to the location of the minimum potential (maximum electron energy) along the punchthrough path. Fig. 8.7 compares the bird's-eye-view of the potential profile for long and short channel MOSFET's. The effect of the drain bias on the potential profile at the channel region is evident for the short channel device.

The I_{DS} vs. V_{DS} curves can provide physical insights into the DIBL behavior of short channel devices. Fig. 8.8 shows the test and simulated data of n-channel devices, with $L_{eff} = 0.5 \mu m$. Log(I_{DS}) is plotted versus V_{DS} for two values of gate to source voltage (V_{GS}), 0 V and 0.2 V. At high drain bias, the drain current increases rapidly and exponentially with the drain voltage in both cases. Also, the drain currents are approximately equal for the two different gate biases. At lower drain bias, the drain current is less dependent on the drain bias, especially for the case with gate bias of 0.2 V. Also, the dependence of the drain current on the gate bias increases. This behavior is due to a change in the punchthrough current path as drain bias increases. The slope is less steep at voltages less than 3 V. This is because the current path is along the surface. The dependence of I_{DS} on V_{DS} is expected to be weaker since the surface potential is mainly controlled by the gate bias. Surface punchthrough occurs before bulk punchthrough for the case of $V_{GS} = 0$ V, due to the surface energy band bending, the combination of channel implants, and shallow junctions, producing a lower potential barrier at the surface. The change from surface punchthrough to bulk punchthrough can be more clearly seen for the case of $V_{GS} = 0.2$ V. The dependence of the drain current on the gate bias is also expected to be large for surface punchthrough, as is confirmed

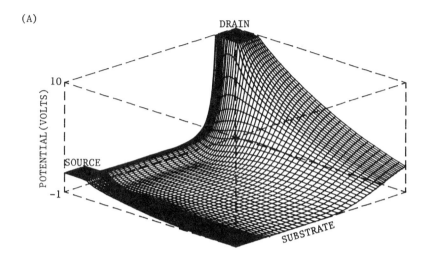

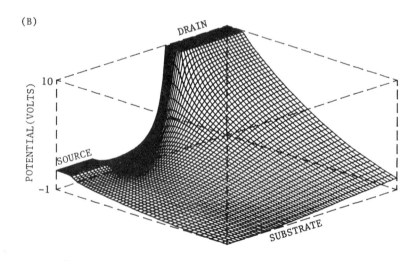

Fig. 8.7. Bird's-eye-view of the potential profile for a long channel (A) and short channel (B) MOSFET's, with a drain and gate bias of 9 and 0 V respectively.

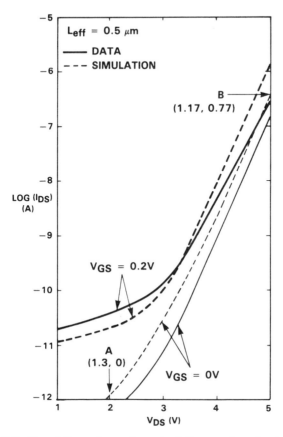

Fig. 8.8. Punchthrough characteristics of short channel MOSFET's, simulation and experimental data.

by the experimental and simulated data. As the drain bias is increased, the barrier in the bulk is lowered, and eventually changes the current path. Since the drain bias has the major effect on the potential in the substrate, the dependence of I_{DS} on V_{DS} is much stronger in this case, as shown by the data with the positive gate bias.

The simulations allow the punchthrough path to be identified. In Fig. 8.8, the punchthrough points (x,y) are indicated for two drain biases for the simulated curve with $V_{GS} = 0$ V. The parameter x is the horizontal distance from the left boundary of the device structure defined in the GEMINI

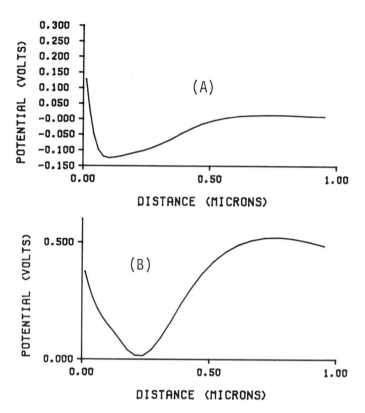

Fig. 8.9. Potential profile vs. vertical distance for points (A) and (B) in Fig. 8.8.

input file. Here the source is located between 0 and 1 μm. The parameter y indicates the distance of the punchthrough point measured from the silicon surface. This point changes from being at the surface to 0.77 μm into the bulk for the drain bias increasing from 2 V to 5 V. This agrees with the previous discussions which concludes that surface punchthrough is occurring at point A and bulk punchthrough is occurring at point B. The potential profile as a function of the vertical distance from the surface into the substrate is plotted in Fig. 8.9, for the two bias points A and B. The horizontal distance from the source is chosen to be equal to the simulated horizontal position of the punchthrough point. The potential maximum (hence energy minimum for electrons) can be seen to be at the surface for bias point A and at the bulk for bias point B. This identifies the surface and bulk punchthrough for the two cases. The punchthrough paths can best be illustrated by the current vectors generated by the PISCES simulations. Fig. 8.10 shows the current vectors for an n-channel transistor with $V_{GS} = 0$ V and $V_{DS} = 3$ V. The L_{eff} is 0.5 μm. The device has no deep boron implant for bulk punchthrough resistance. The current vectors indicate that current is flowing along the surface and through the bulk. The bulk punchthrough path has a maximum depth of about 0.6 μm from the silicon surface. A boron implant of 70 KeV and a dose of about 1E12 cm^{-2} will eliminate the bulk path.

Another interesting punchthrough phenomenon is shown in Fig. 8.11 for a p-channel transistor with n$^+$ polysilicon gate. The transistor has a counter-doping in the channel for threshold voltage adjustment (see Ch. 12). In this case, the punchthrough path is through the entire counter-doping layer, with a layer thickness of about 0.2 μm. This result indicates that the counter-doping causes serious punchthrough problem at short channel length. This is a major concern for submicron CMOS using n$^+$ polysilicon gate.

Overall the simulations agree with the experiment very well qualitatively. To agree quantitatively well with experimental data, one needs very precise process simulations of the channel and source/drain profile, and also very accurate experimental data such as the effective channel length. Therefore, very good quantitative agreements for short channel devices at high drain biases are difficult to obtain. But still one can gain a lot of insight and have good estimates on the punchthrough behavior of submicron channel length MOSFET's.

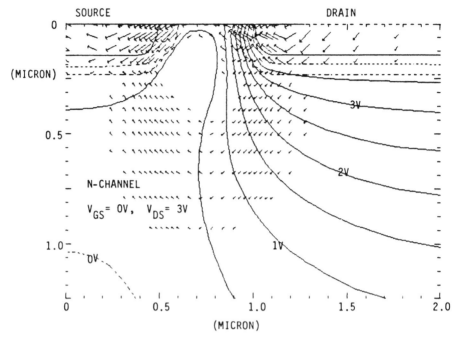

Fig. 8.10. Simulation of bulk punchthrough in n-channel MOSFET's using PISCES.

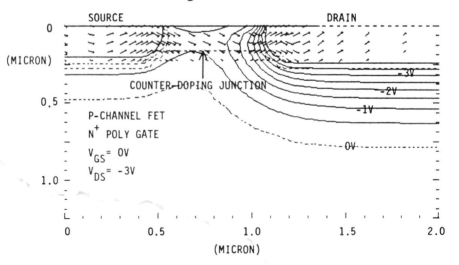

Fig. 8.11. Simulation of the punchthrough characteristics of a p-channel MOSFET's using PISCES.

References

[8.1] R. R. Troutman, "VLSI Limitations from Drain-Induced Barrier Lowering," *IEEE Trans. Electron Devices*, **ED-27**, April 1979, pp. 461-468.

[8.2] R. R. Troutman, "Subthreshold Design Considerations for Insulated Field Effect Transistors," *IEEE J. Solid State Circuits*, **SC-9**, April 1974, pp. 55-60.

[8.3] B. Eitan and D. Frohman-Bentchkowsky, "Surface Conduction in Short Channel MOS Devices as a Limitation to VLSI Scaling," *IEEE Trans. Electron Devices*, **ED-26**, April 1979, pp. 254-266

[8.4] H. Masuda, M. Nakai, and M. Kubo, "Characteristics and Limitation of Scaled-Down MOSFET's Due to Two-Dimensional Field Effect," *IEEE Trans. Electron Devices*, **ED-26**, June 1979, pp. 980-986.

[8.5] K. M. Cham and S. Y. Chiang, "Device Design for the Submicrometer P-Channel FET with n^+ Polysilicon Gate," *IEEE Trans. Electron Devices*, **ED-31**, July 1984, pp. 964-968.

[8.6] J. J. Barnes, K. Shimohigashi, and R. Dutton, "Short-Channel MOSFET's in the Punchthrough Current Mode," *IEEE Trans. Electron Devices*, **ED-26**, April 1979, pp. 446-453

[8.7] Y. A. El-Mansy and R. A. Burghard, "Design Parameters of the Hi-C DRAM Cell," *IEEE J. Solid-State Circuits*, **SC-17**, Oct 1982, pp. 951-956.

[8.8] H. Katto, K. Okuyama, S. Megure, R. Nagai and S. Ikeda, "Hot Carrier Degradation Modes and Optimization of LDD MOSFET's," *Tech. Digest of IEDM 1984*, pp. 774-777.

[8.9] S. Odanaka, M. Fukumoto, G. Fuse, M. Sasago, T. Yabu, and T. Ohzone, "A New Half-Micrometer P-Channel MOSFET's with Efficient Punchthrough Stops," *IEEE Trans. Electron Devices*, **ED-33**, Mar. 1986, pp. 317-321.

Chapter 9
A Study of LDD Device Structure Using 2-D Simulations

In this chapter, analysis and design of LDD (Lightly Doped Drain) devices using two-dimensional device simulation and experiments will be described to illustrate the usefulness and necessity of using computer-aided design tools in the fabrication of VLSI devices. First, the problem of high electric field in VLSI devices and the use of LDD device as a possible solution is discussed. The fabrication and simulation of LDD device is then described. Finally, the performance, characteristic, physics and design considerations of LDD device are presented in detail.

9.1 High Electric Field Problem in Submicron MOS Devices

In Very Large Scale Integrated (VLSI) circuits, the dimensions of devices are continually being scaled down to obtain higher density and speed. The channel length, junction depth and gate oxide thickness are scaled down while the channel doping is scaled up. These scalings are necessary to provide more current drive while maintaining the right device threshold voltage and low leakage current. However, the power supply voltage tends to remain unchanged due to system requirements. The same voltage is hence being applied to a much shorter channel which results in a higher channel electric field. In addition, the shallower junction and thinner gate oxide also make the electric field at the drain junction higher. Eq. (9.1a,b) [9.1],[9.2] is a simple analytical expression relating the maximum drain electric field to device parameters.

$$E_m = \frac{(V_D - V_{DSAT})}{\sqrt{3T_{ox}X_j}} \tag{9.1a}$$

$$V_{DSAT} = \frac{(V_G - V_T)LE_{SAT}}{(V_G - V_T + LE_{SAT})} \tag{9.1b}$$

where E_m is the maximum drain electric field, V_D is the applied drain voltage, V_{DSAT} is the drain saturation voltage, T_{ox} is the gate oxide thickness, X_j is the junction depth, L is the effective channel length and E_{SAT} is the critical field for velocity saturation and is about 3E4 V/cm.

It can be seen that as gate oxide thickness, junction depth and channel length are scaled down, the maximum drain electric field increases. Electrons in the channel are accelerated by this high drain electric field to reach high energy. These high energy electrons are called 'hot electrons' and they can cause device failure and long term reliability problems.

One problem is that some electrons can reach high enough energy to cross over the silicon/silicon dioxide barrier and create damage at the oxide/silicon interface. The interface damage can cause threshold voltage shift and also reduce the mobility of electrons. The current driving capability of the device is hence continually being degraded during device operation. This presents a severe reliability problem because circuit failure can occur after a certain period of operation. Another problem is that the high energy electrons can cause impact ionization in the drain depletion region, generating electron/hole pairs. The electrons are collected by the drain while the holes go into the substrate. The substrate hole current forward biases the source junction and electrons are injected into the substrate. These electrons will be collected at the drain as excess current and generate more impact ionization substrate current. This is a positive feedback mechanism which can cause the drain current to increase rapidly. The resulting avalanche breakdown will destroy the device. Excessive substrate current can also cause problems to substrate bias generation circuits and is a potential cause of latchup in CMOS circuits. Eq. (9.2) [9.3] is an empirical analytical expression relating substrate current to maximum drain electric field.

$$I_{sub} \approx 2I_{DS}\exp(\frac{-1.7E6}{E_m}) \quad (9.2)$$

where I_{DS} is the drain current and E_m is the maximum drain field.

New device structures are needed to reduce the drain electric field and hence prevent hot electron degradation and catastrophic drain breakdown. Several device structures have been proposed, such as lightly doped drain (LDD) device [9.4],[9.5] and As/P double diffused device [9.6]. The idea is to use a lightly doped region to drop off some drain voltage so that the drain electric field can be lowered. A more graded junction also helps to reduce drain electric field. In the design of such devices, a lot of issues need to be considered. Major concerns are whether the additional n⁻ region will degrade device performance, what is the right doping concentration of the n⁻ region to achieve both low electric field and acceptable series resistance, and whether low electric field leads to less hot electron related problems. In this chapter, only the LDD device will be discussed.

9.2 LDD Device Study

Lightly doped drain (LDD) device has received much attention in recent years as an important VLSI device. The major advantage of LDD device is that the lightly doped region can reduce the peak electric field at the drain. This results in higher breakdown voltage, lower substrate and hot electron currents and hence improved reliability. Other proposed advantages are reduction of short channel effect and improvement in punchthrough voltage due to the shallower tip junction. Experimental results on LDD device have been reported in several papers [9.4],[9.5]. However, the physics of submicron LDD device is not well understood. This is due to the complexity of two-dimensional device effects and difficulties in measuring and extracting parameters for submicron devices. In this section, the device physics for a half micron LDD device is investigated by using the two-dimensional device simulation program CADDET together with experimental results. The combination of computer simulation and experimental analysis is found to provide important insights into the understanding of the LDD device in terms of breakdown, trapping, punchthrough and others. Simulation results are found to agree well with experiments.

Device Fabrication and Simulation

LDD device differs from conventional devices by having a n^- region at both the source and drain side. Fig. 9.1 shows the conventional and LDD device respectively. Different processing methods can be used to create the LDD device structure. In our laboratory, a n^- implant is done after the polysilicon gate is defined and etched. (The n^- implant dose will be in unit of cm^{-2} throughout the chapter.) A thin oxide is grown after the implant and then a layer of oxide from 100 nm to 300 nm is deposited by low pressure chemical vapor deposition (LPCVD) at $900°$ C. This oxide layer is then anisotropically etched to form an oxide spacer at the polysilicon gate edge. Heavy dose Arsenic implant is then done to create the n^+ source and drain junctions. Fig. 9.2 shows the processing sequence used to fabricate the LDD device. The device used in this work has gate oxide thickness of about 20 nm, a junction depth of about 0.25 micron and a n^- region with various doping levels and lengths.

The two-dimensional simulation program CADDET used to study the LDD device solves both the Poisson equation and the current continuity equation simultaneously. The program does not have an impact ionization model and hence cannot simulate substrate current. However, since the substrate current is related to the maximum drain electric field (Eq. (9.2)), very useful information can still be obtained. In CADDET, the n^- region can only be placed either on the source or drain end. The tip is hence placed on the source side to study the effect of series resistance on device characteristics and on the drain side to study drain electric field effects. On the source side, the gate always covers the entire source region, while on the drain side, the overlap between gate and drain can be varied. Both the length of the n^- region and its doping profile can be varied. The particular limitations of CADDET need to be taken into account in the interpretation of simulation results. Fig. 9.3 shows

LDD Device

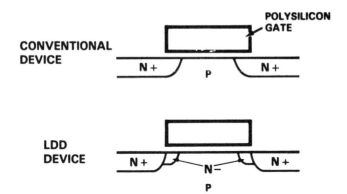

Fig. 9.1. Conventional and LDD device structures.

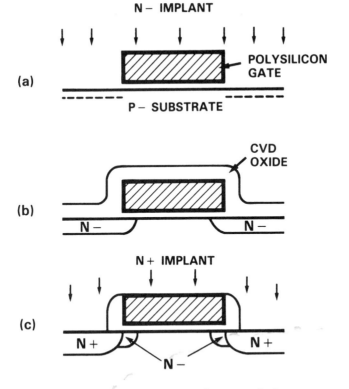

Fig. 9.2. Processing sequence for LDD device.

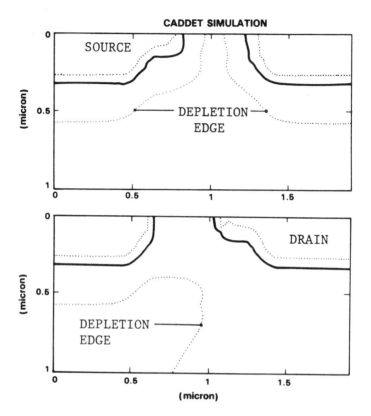

Fig. 9.3. Structures used for source and drain simulations.

the structure used in simulating the source and drain side respectively. Note that the device channel widths used in the simulation is different from that in the experimental data. Therefore only the trend should be compared.

Performance Comparison Between LDD and Conventional Device

One common criterion used to evaluate a certain device technology is to plot the saturation current versus the punchthrough voltage for the devices. The idea is that the shorter the channel length, the higher is the current but at the same time, the lower is the punchthrough voltage. For the same channel length, a technology with a deeper junction will have lower punchthrough voltage while one with lower source/drain doping and shallower junction depth (hence higher series resistance), will have lower saturation current. The saturation current versus punchthrough voltage characteristic tells us the best current drive we can get subject to a certain punchthrough voltage requirement and is hence a good evaluation of the performance of a technology. In simulating the performance comparison between LDD and conventional device, a 0.2 micron n⁻ tip is placed at the source side. The n⁻ dose and channel lengths are varied. For punchthrough simulation, the gate voltage is set at zero volt and the drain voltage is stepped in 0.1 volt increments. The punchthrough voltage is defined as the drain voltage at which the drain current is 10 nA per micron channel width. Saturation current is defined at $V_G = V_D = 5$ V. Fig. 9.4 shows a plot of the simulated saturation current versus punchthrough voltage characteristic. It can be seen that for the same punchthrough voltage, the saturation current of the LDD device is about the same or lower than conventional devices, depending on the dose of the n⁻ region. Fig. 9.5 is a plot of I_{DSAT} versus punchthrough characteristic from some experimental LDD and conventional devices, which agrees with the simulation. The reason for such inferior performance of LDD in the half micron device is due to the fact that the series resistance of the n⁻ region is quite comparable to the channel resistance. The gain in punchthrough voltage due to the shallow n⁻ junction is more than offset by the decrease in current due to the series resistance. Fig. 9.6 and 9.7 plots the simulated linear and saturation transconductance characteristic for conventional and LDD devices under different gate bias. The numbers in parenthesis are the effective channel lengths. In the linear region, the percentage degradation becomes worse with increasing gate voltage, whereas in the saturation region, degradation effect is less and decreases with increasing gate voltage. The experimental linear and saturation transconductance characteristics shown in Fig. 9.8 and 9.9 show the same trend. Since in CADDET, the n⁻ region on the source side is entirely covered by the gate, the effective series resistance is reduced at high gate voltage and the degradation due to this resistance is being under estimated. However, CADDET does provide the right direction and saves a lot of time and effort in experimentation. The typical time required to fabricate and test the devices is about 2-3 months while the simulations can be done on a HP1000 minicomputer in a few days. Moreover,

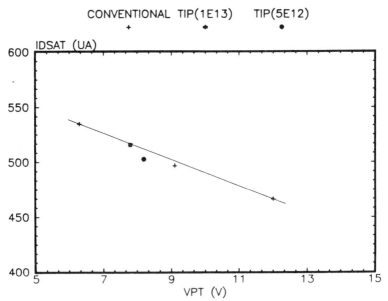

Fig. 9.4. Saturation current vs. punchthrough voltage, simulation. Tip length = 0.2 μm.

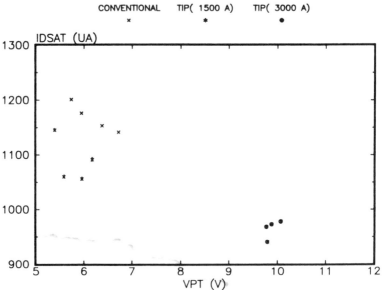

Fig. 9.5. Saturation current vs. punchthrough voltage, experimental data. Tip dose = 5E12 cm^{-2}.

LDD Device

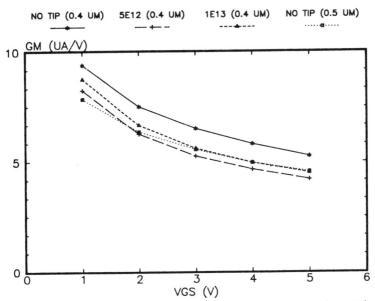

Fig. 9.6. Simulated linear transconductance vs. gate bias. Tip length = 0.2 μm.

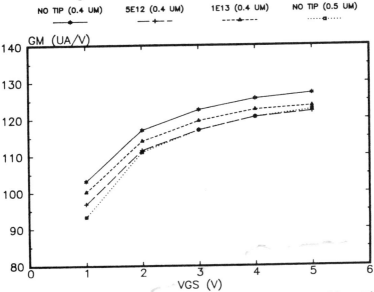

Fig. 9.7. Simulated saturation transconductance vs. gate bias. Tip length = 0.2 μm.

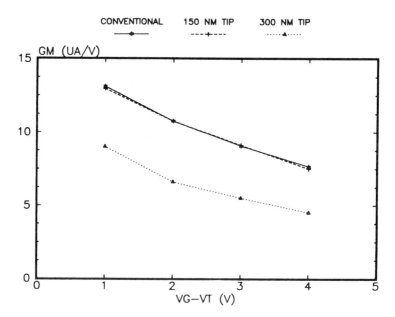

Fig. 9.8. Experimental linear transconductance vs. gate bias. Tip dose = 5E12 cm^{-2}.

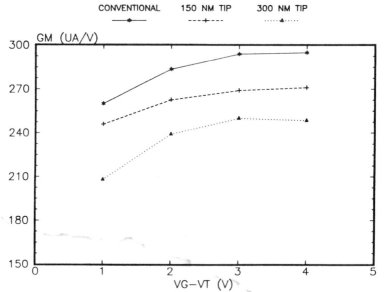

Fig. 9.9. Experimental saturation transconductance vs. gate bias. Tip dose = 5E12 cm^{-2}.

LDD Device

the physics of the effects can be readily investigated by looking at potential, electric field and electron distributions in the simulated device.

Drain Electric Field and Substrate Current Study

The drain electric field characteristic of LDD device is simulated by putting the n⁻ region at the drain side. The drain electric field as a function of n⁻ region doping, n⁻ region length and the amount of gate overlap of the n⁻ region is studied. Several interesting results are obtained through the simulation and are confirmed experimentally by looking at the substrate current characteristics. Important physical insight into the operation and design of LDD device, conventional device and minimum overlap devices are obtained.

A) *Drain electric field as a function of* n⁻ *region dose*

In the simulation of drain electric field characteristic as a function of n⁻ region doping, it is found that, contrary to common belief, lower n⁻ region doping doesn't necessarily lead to lower drain field. Fig. 9.10 plots the simulated drain peak field versus n⁻ region implant dose for n⁻ region length of 0.2 micron and a bias condition of 5 V at the drain and 3 V at the gate. A minimum peak field is found at a dose of 5E12. Physically, for the lowly doped case (2E12), the entire n⁻ region is depleted and a high field occurs at the high-low junction. For the higher doping (1E13), it is harder to deplete the n⁻ region and the peak field occurs at the edge of the n⁻ region. Experimental results confirm this finding. Fig. 9.11 plots the drain voltage needed to generate $1\,\mu A$ per micron of channel width of substrate current versus the n⁻ region implant dose. It can be seen that the highest drain voltage required is at a dose of 5E12. This optimum dose for breakdown and lowest substrate current will vary to a certain extent with the n⁻ region length. Although 5E12 gives the best breakdown and substrate current characteristics, the series resistance is high as shown in Figures 9.4 to 9.9. Another consideration is that lower substrate current does not necessarily lead to lower hot electron trapping. All these factors need to be considered in the optimum design of the device.

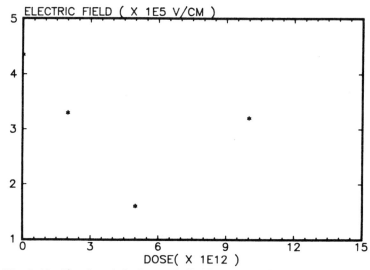

Fig. 9.10. Simulated drain peak field vs. n⁻ region implant dose. Tip length = 0.2 μm.

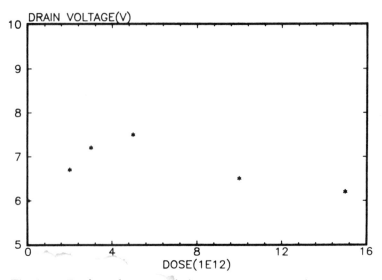

Fig. 9.11. Drain voltage needed to generate 1 μA/μm channel width of substrate current vs. n⁻ region implant dose, experimental data. Tip length = 0.2 μm.

LDD Device

B) *Drain electric field versus gate voltage characteristics*

According to Eq. (9.1a), in the conventional device the maximum drain electric field is directly proportional to $V_D\text{-}V_{DSAT}$. Since V_{DSAT} increases with V_G (Eq. (9.1b)), the drain electric field decreases with increasing gate voltage. From Eq. (9.2), reduction in drain electric field means the substrate current will also decrease with increasing gate voltage. Fig. 9.12 shows the measured substrate current versus gate voltage characteristic of a conventional device. It can be seen that after an initial peak, the substrate current does decrease with increasing gate voltage. The initial increase of substrate current with gate voltage is due to increase in channel current.

The drain electric field versus gate voltage characteristic is simulated for LDD structures with different n⁻ region doping and amount of overlap between the gate and the n⁻ region. The electric field at the surface along the channel is plotted for each bias point and the position where the peak electric field occurs is noted. Fig. 9.13 shows a plot of the simulated electric field along the channel for the bias point of $V_G = 3$ V and $V_D = 6$ V. The electron concentration as a function of Y (the distance from silicon/silicon dioxide interface into silicon) is then plotted at the position of the channel where peak electric field occurs. The position where peak electron concentration occurs is noted. Depending on the device structure, the peak electron concentration in some cases is not at the surface. The electric field along the channel is then plotted again at the Y position where peak electron concentration occurs. Then the maximum drain electric field is found from this plot. Fig. 9.14 is a plot of maximum drain electric field as a function of gate bias for different amount of gate overlap at a n⁻ dose of 1E13. It can be seen that for low V_G, the drain electric field is greatly reduced when the gate overlaps the n⁻ region more. Another observation is that the drain electric field is actually not a very sensitive function of gate voltage for half micron LDD device for X larger or equal to 0.2 μm. Fig. 9.15 is a similar plot for a n⁻ region dose of 5E12. The characteristic is very different from that of Fig. 9.14. It can be seen that the electric field initially decreases with increasing gate voltage, reaches a minimum and then goes up again. The smaller the gap between the gate and the n⁺ junction, the higher the electric field goes up. This effect was confirmed in experimental devices. Fig. 9.16

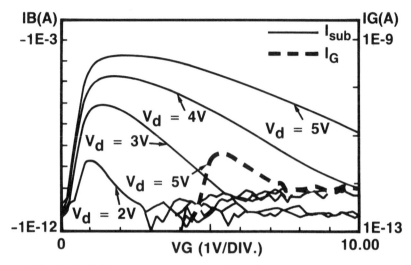

Fig. 9.12. Substrate current vs. gate bias for conventional device.

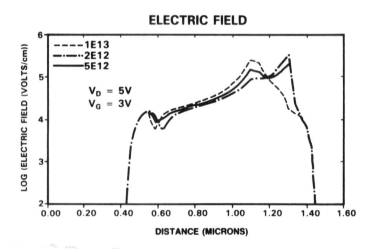

Fig. 9.13. Simulated electric field along the channel.

LDD Device

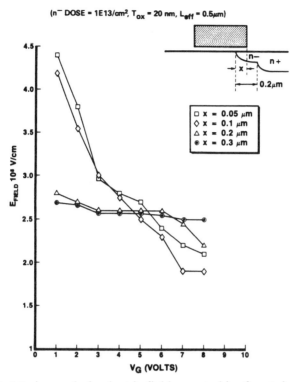

Fig. 9.14. Maximum drain electric field vs. gate bias for n⁻ dose of 1E13.

shows substrate and gate current versus gate voltage characteristics for LDD devices with a n⁻ region doping of 5E12 and different n⁻ region length. A double hump is observed in the substrate current characteristics which correspond to a decreasing and then increasing electric field.

The physics of this phenomenon can be understood by looking at the potential, electric field and carrier density along the channel. Fig. 9.17 plots the channel potential and electric field for $V_G = 2$ and 8 V. It can be seen that at $V_G = 2$ V, most of the drain potential is dropped in the n⁻ region overlapped by the gate (X) and the peak electric field occurs at the edge of the n⁻ region. For $V_G = 8$ V, most of the drain potential is dropped in the region between the gate edge and the edge of the n⁺ junction (Y) and the peak electric field occurs

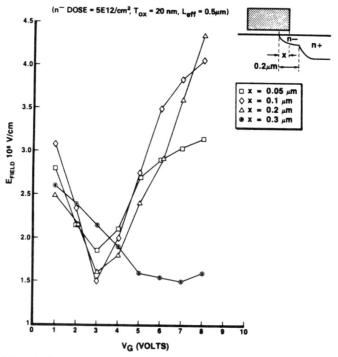

Fig. 9.15. Maximum drain electric field vs. gate bias for n⁻ dose of 5E12.

at the Hi-Lo junction. In the latter case (V_G = 8 V), the shorter the distance Y, the higher is the electric field because the potential is dropped in a shorter distance. The way the potential and electric field behave under different gate bias conditions is found to be due to the fact that at high gate bias the channel carrier density can be comparable to the doping concentration of the n⁻ region. At V_G = 2 V, the channel carrier density is less than the n⁻ region doping concentration. The n⁻ region under the gate is mostly depleted and most of the drain potential is dropped there. The peak electric field occurs at the edge of the n⁻ region. However, at V_G = 8 V, the channel carrier density is comparable to the n⁻ region doping concentration (for the 5E12 case), therefore the n⁻ region under the gate can no longer be considered depleted. The space charge region is pushed towards the heavily doped n⁺ junction and the potential is then dropped in the region Y. The position of the

LDD Device

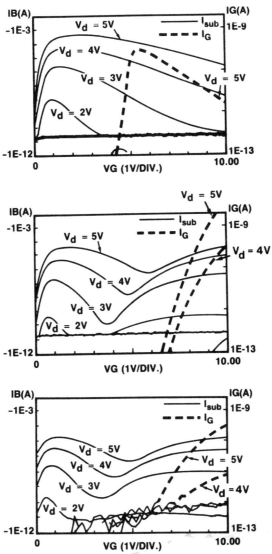

Fig. 9.16. Measured substrate and gate current vs. gate bias for LDD devices with different spacer widths. Top: 120 nm, Middle: 200 nm, Bottom: 300 nm. (All with n^- doping = 5E12, T_{ox} = 20 nm, W/L = 100/1).

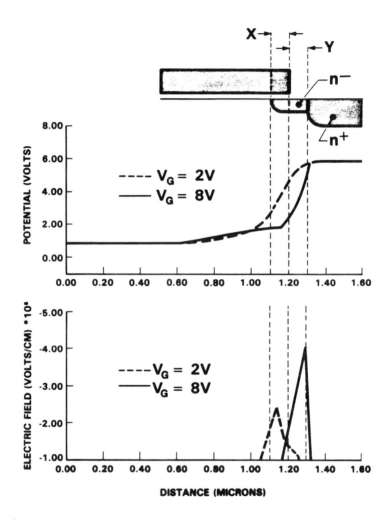

Fig. 9.17. Channel potential and electric field along the channel for V_G = 2 and 8 V.

peak electric field also moves from the p/n⁻ junction to the n⁻/n⁺ junction. The narrower the region Y is, or the higher the channel current, the higher is the drain electric field which results in higher substrate current. In this respect, this phenomenon resembles the Kirk effect in bipolar transistors. Since the effect is due to channel carrier density being comparable to n⁻ region doping concentration, increasing n⁻ region doping will suppress this effect while reducing gate oxide thickness will enhance it. This double hump substrate current characteristic is found in minimum overlap devices. When there is not enough overlap between the n⁺ source/drain and the gate, a graded n⁻ region exists between the edge of the gate and the n⁺ region and this phenomenon can be observed. This characteristic can hence be used as a tool to check if there is enough overlap between the source/drain and the gate. Adequate overlap between the source/drain and gate is important both for device performance and reliability.

9.3 Summary

Computer simulation together with experiments provide a good understanding of the physics of LDD devices. Submicron LDD device is found to provide lower drain electric field at the expense of lower current drive. A n⁻ region doping of 5E12 is found to give the lowest drain electric field although this may not be optimum from the standpoint of series resistance and hot electron damage effects. A dose of 1E13 is probably a good compromise. A new substrate current characteristic is found and explained. This can be used as a tool to check for inadequate overlap between source/drain and gate. This physical understanding help us to be aware of the various tradeoffs in the design of an optimum LDD device for a certain circuit application.

In this study, computer simulation is found to have several advantages. First of all, the physics of the device can be investigated in detail. Potential distribution, electric field and electron concentrations can be plotted out easily and analyzed. This is essential for submicron device study because of the complicated two-dimensional nature and other subtle effects. Without the help of the simulation, it would take much more time and effort to understand the double hump phenomenon in the substrate current characteristic. Another important feature is the ease of varying different device parameters. This saves a lot of time and money and at the same time eliminates uncertainties in experiments. The device structure is well defined in terms of junction doping profile, channel length and channel doping profile. However, a simulator is only as good as the physical models in it. One must be aware of the limitations in the physical models, device structures and other specifications in the model and interpret the results properly. The simulation results need to be confirmed with experiments. Simulators can be used to guide us in the direction of more intelligent experiments and greatly reduce the number of experiments needed. Experimental results can also be interpreted with the help of simulations. This study on LDD devices shows that proper use of simulation with experiment can be very effective in the understanding and design of submicron devices.

References

[9.1] Cheming Hu, "Hot electron effects in MOSFET's", *Tech. Digest of IEDM 1983*, pp. 176-179.

[9.2] P. K. Ko, Ph.D. thesis, University of California, Berkeley, CA 1982.

[9.3] S. Tam, P.-K. Ko, C. Hu and R. S. Muller, "Correlation between Substrate and Gate Current in MOSFET's," *IEEE Trans. Electron Devices*, **ED-29**, pp. 1740-1744, Nov 1982.

[9.4] P. J. Tsang, S. Ogura, W. W. Walker, J. F. Shepard and D. L. Critchlow, "Fabrication of High performance LDDFET's with oxide sidewall-spacer technology," *IEEE Trans. on Electron Devices*, **ED-29**, no.4 pp. 590-595, April 1982.

[9.5] Seiki Ogura, Christopher F. Codella, Nivo Rovedo, Joseph F. Shepard and Jacob Riseman, "A half micron MOSFET using double implanted LDD," *Technical Digest of IEDM 1982*, pp. 718-721.

[9.6] K. Balasubramanyam, M. J. Hargrove, H. I. Hanafi, M. S. Lin, D. Hoyniak, J. LaRue and D. R. Thomas, "Characterization of As-P Double Diffused Drain Structure," *Tech. Digest of IEDM 1984*, pp. 782-785.

Chapter 10

The Surface Inversion Problem in Trench Isolated CMOS

10.1 Introduction to Trench Isolation in CMOS

Trench isolation has recently been proposed for advanced CMOS processes [10.1]-[10.4]. Fig. 10.1 shows a trench isolated CMOS structure. The major advantage of trench isolation in CMOS is to reduce the latchup problem [10.5],[10.6]. Fig. 10.2 shows the latchup path for a CMOS structure, where the parasitic bipolar transistors are shown to form a positive feedback path. If there is excess current flow in the n-well and substrate, and if the substrate and n-well resistances are large enough, significant voltage drops will occur in the substrate and n-well. Under this condition, the parasitic bipolar transistors can be turned on [10.5]. This phenomenon has been studied by many researchers in CMOS technology. The trench has been suggested as a good candidate for isolation between n-channel and p-channel transistors. For an n-well process it is expected that the lateral n-p-n parasitic bipolar transistor gain will be reduced, thereby increasing the latchup initiating current. In a p-well process, the reduction of latchup sensitivity is achieved from reduction of the lateral parasitic p-n-p current gain. Besides providing higher resistance to latchup, the trench also reduces the lateral side diffusion of the well, hence

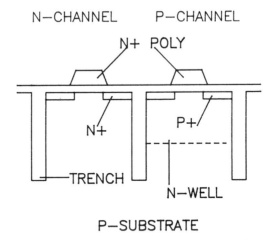

Fig. 10.1. Trench isolation in CMOS technology.

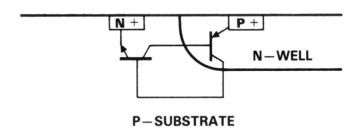

Fig. 10.2. Latchup path in CMOS.

Trench Isolation

allowing a higher packing density.

At present, trench isolation processes use a combination of trench isolation and a field isolation, typically LOCOS [10.4]. This is necessary to avoid leakage problems at the trench surface due to inversion, and also provide field area for interconnections. For our discussion, we assume that the trench is etched with vertical walls and refilled with oxide in a processing sequence that is consistent with low defect VLSI technology. It is interesting to see if the transistors can be placed adjacent to the trench side-wall. If this is possible, then it will provide the highest packing density for bulk CMOS technology, if the interconnection process is also optimized. Also, the narrow width effects on the transistors (see Chapter 11) are eliminated because the trench has a vertical wall, in contrast to the conventional LOCOS process. For LOCOS, the bird's beak has caused the loss of channel width in narrow width transistors, and also causes the increase of threshold voltage with decreasing channel width. Since trench isolation has no field implant, there is no boron encroachment into the active area.

It would be possible to place the transistors next to the trench if the trench surface is ideal. However, this may not be the case. The trench surface may be inverted due to positive charges at the trench surface or in the trench refill material [10.7],[10.8]. The effect of the plasma used to etch the trench may enhance the charge density at the trench surface. This problem is more serious for n-well CMOS because the n-well is biased at a positive voltage. The n-well in this case acts like an electrode which tends to invert the p-type substrate region opposite to the n-well. Also, the substrate doping concentration decreases with increasing distances into the substrate. This means that a relatively low charge density at the surface of the trench may invert the trench surface. If the trench surface is inverted, then there will be a conducting path connecting the n-well and the n^+ region. This increases the effective area of the collector of the lateral npn parasitic bipolar transistor, hence causing latchup problems. The inversion will also cause leakage between isolated n^+ regions and between the source and drain of the n-channel transistor. The problem of trench surface inversion has been studied using simulations to identify the problem under different processing and bias conditions. From the simulation results, appropriate application of the trench isolation is recommended. Experimental results are also presented.

10.2 Simulation Techniques

Before doing the numerical simulations, it is useful to roughly estimate the range of fixed charge density (Q_{ss}) which would cause inversion at the trench surface. A simple model would be an MOS capacitor with the n-well as the gate, the trench as the dielectric with the dielectric thickness equal to the trench width. Positive charge with density of Q_{ss} cm^{-2} is assumed to be at the trench surface. Fig. 10.3 shows the structure. Then the threshold voltage is simply given by equation [10.9]:

$$V_T = \phi_{ms} + \frac{Q_B - Q_{ss}}{C_{ox}} + 2\psi_B \qquad (10.1)$$

where the first term is due to the work function difference between the n-well and p-substrate, the second term due to the depletion charge density at inversion (Q_B) and trench surface charge density, and the last term is the surface potential bending necessary to achieve inversion. Here C_{ox} is the dielectric capacitance between the trench capacitor. The values of ϕ_{ms}, Q_B, and ψ_B can be easily calculated from the doping concentration of the n-well and p-substrate.

The threshold as a function of Q_{ss} for a p-substrate doping of 1E15 and 1E16 cm^{-3} and n-well doping of 1E16 cm^{-3} is shown in Fig. 10.4. It is clear that Q_{ss} is a problem for trench isolation for lightly doped substrate.

In order to investigate this problem in more detail, taking into account the effect of a varying p-type impurity distribution along the trench surface, two-dimensional simulations are performed. Also, the simulations can provide information about the region where inversion is most likely to occur.

The trench structure can be generated using the GEMINI program which solves Poisson's equation. The only caution in using this program is that the quasi-Fermi level along the channel is set equal to the drain quasi-Fermi level, even after inversion. This means that the results are valid only at or before inversion occurs at the trench surface. After inversion has occurred, the potential value at the trench surface would be overestimated, and so would be the extension of the inverted region.

Trench Isolation

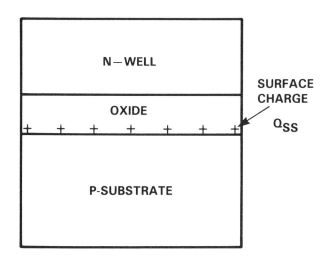

Fig. 10.3. Capacitor for trench surface inversion calculation. The oxide thickness is equal to the trench width.

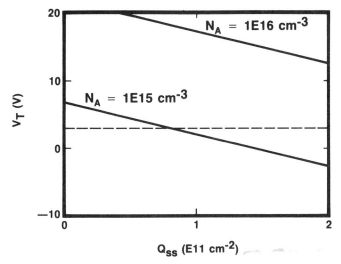

Fig. 10.4. Trench surface threshold voltage vs. Q_{ss} and p-type substrate concentration. The dotted line indicates a threshold voltage of 3 V.

The structure is generated using the MOSFET structure, with the thickness of the gate oxide equal to the trench depth. Then the source and drain regions are raised to the top of the trench. The drain of the MOSFET becomes the n-well, and the source becomes an n^+ region. The impurity profile can either be generated by the SUPREM program, or can be specified as Gaussian profiles in the GEMINI input file. The maximum number of grids in the vertical direction is 99, therefore, the trench depth is set to be 5 μm, and n-well to be 3 μm, such that a realistic structure is simulated with good accuracy. If the trench is made much deeper, the grid spacing would be too large. The number of iterations for convergence is typically 300, compared with values of 100 or less for conventional MOSFETs. Fig. 10.5 shows the generated structure. Also specified in the GEMINI input file are the applied bias at the n-well, n^+, and also at the substrate. The contacts to the n^+, n-well and substrate are assumed to exist and will not be shown. The Q_{ss} values are specified in the input file, and set to be a constant along the oxide-silicon interface. Fig. 10.6 shows the bird's-eye-view of the impurity profile for the trench structure. The n^+ region, n-well, and p-substrate are shown. Also shown is the p-type channel implant which overlaps the n^+ profile. This channel implant is used to adjust the threshold voltage of the n-channel transistor. Because of the higher p-type doping concentration near the n^+ region, it will be shown that this reduces the probability of inversion near that region. The impurity concentration within the trench should be neglected in the interpretation of this plot. (The PLPKG program sets the impurity concentration within the trench equal to the substrate concentration at the bottom of the trench.)

10.3 Analysis of the Inversion Problem

In this section, the interpretation of the simulated potential distribution will be discussed. Fig. 10.7 shows a typical two-dimensional plot of the trench simulation. The constant potential contours are shown, as well as the depletion regions and p/n junctions. The potential is equal to the conduction band potential in the neutral substrate, and is set equal to the value of the applied substrate bias voltage in the neutral substrate region, i.e., outside the depletion

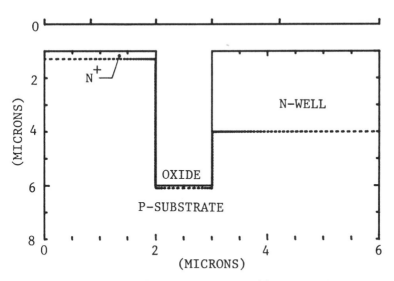

Fig. 10.5. Trench structure generated by GEMINI.

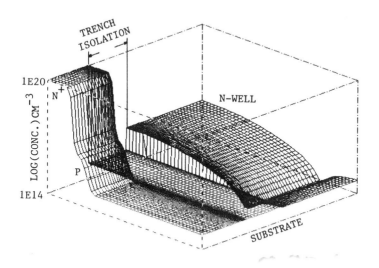

Fig. 10.6. Bird's-eye-view of the impurity profile.

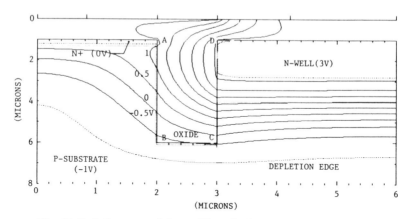

Fig. 10.7. 2-D potential profile of the trench isolated CMOS structure. The n-well bias is 3 V.

region. The bias conditions are 0, 3, -1 volts at the n^+, n-well and substrate respectively. The p-type channel implant profile which overlaps the n^+ region has been neglected in this case for simplicity. The shape of the potential contours indicates that the n-well and Q_{ss} both have the effect of raising the potential at the trench surface opposite to the n-well. It is interesting to look at the potential along the trench surface from the n^+ to the n-well, as follows:

The path along the trench surface from the n^+ to the n-well can be viewed as the channel of a MOSFET with the n^+ as the source, the n-well as the gate and the drain, although part of the channel is not under the gate. From the results of the two-dimensional simulations, the potential along this path can be plotted. Fig. 10.8 shows the potential profile for different cases. The first case is for no surface charge and zero n-well bias, as a reference. The second case shows the effect of increasing the n-well bias to 3 V. The potential in the region opposite to the n-well is raised, as expected. The approximate quasi-Fermi level is also shown. The quasi-Fermi level is a function of the biases at the n^+ and n-well. The condition for inversion is similar to the gated diode structure [10.10]. If the potential is above the quasi-Fermi level, then inversion occurs. In this case, it is clear that no inversion has occurred, as expected. The third case is for a positive charge density of 5E10 cm^{-2} at the trench surface, and with an n-well bias of 3 V. The potential near the n^+

Trench Isolation 241

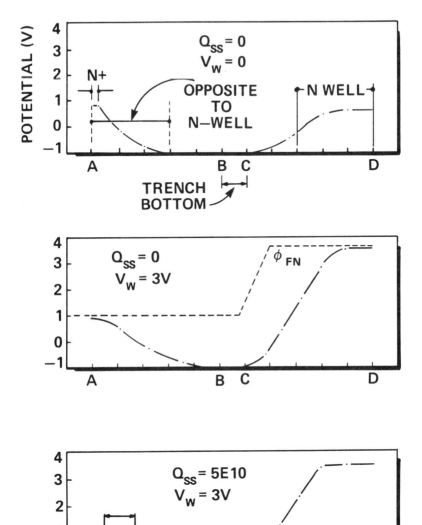

Fig. 10.8. Simulated trench surface potential from the n^+ region to the n-well, for different Q_{ss} and n-well bias.

region and opposite to the n-well is above the quasi-Fermi level at the source, hence inversion will occur in this region, as indicated in the figure. Fig. 10.9 shows the bird's-eye-view of the potential profile. The potential is set equal to the substrate bias voltage in the neutral substrate region. Therefore the n^+ potential and n-well potential are equal to their applied bias added to the built-in potential of the p/n junction. The potential "bump" near the source and opposite to the n-well is due to the combined effect of the positive surface charge and n-well bias. Inversion occurs when the trench surface potential is higher than $2\psi_B$. Qualitatively, in the bird's-eye-view plot of the potential, if the height of the bump is higher than the n^+ potential, inversion will occur. In this case, inversion has occurred.

The effect of the p-type channel implant on the inversion problem is also studied. Fig. 10.10 shows the vertical impurity profile at the n^+ region. The p-type channel implant has a significant effect on the trench surface inversion near the n^+ region, since it increases the p-type impurity concentration near that region. Fig. 10.11 shows the GEMINI simulation where the p-type channel implant which overlaps the n^+ region is taken into account. The two-dimensional profile shows a reduction of the potential in that region. In this case, the probability of inversion is reduced, under the same conditions. Fig. 10.12 shows the bird's-eye-view of the potential profile. The magnitude of the potential "bump" is reduced, in comparison to Fig. 10.9. The potential is also suppressed at the edge of the n^+ region due to the higher boron concentration. Note that this channel implant will be increased in scaled down devices to prevent drain-induced barrier lowering problems, as well as to adjust the threshold voltage for thinner gate oxide. Therefore, the inversion problem near the n^+ region will be reduced for scaled CMOS. The trench surface inversion in the bulk, however, will be dependent only on the charge density and the substrate doping concentration.

10.4 Summary of Simulation Results

The trench simulation has been performed for different values of Q_{ss} and bulk doping concentration, n-well bias, and substrate bias. The value of Q_{ss}

Trench Isolation

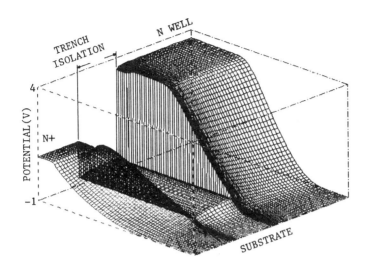

Fig. 10.9. Bird's-eye-view of the potential profile.

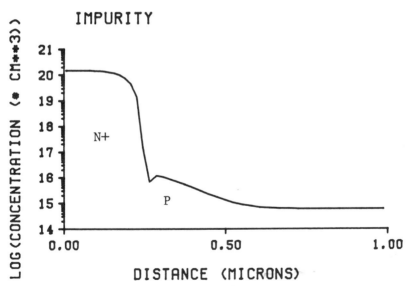

Fig. 10.10. Vertical impurity profile at the n^+ region.

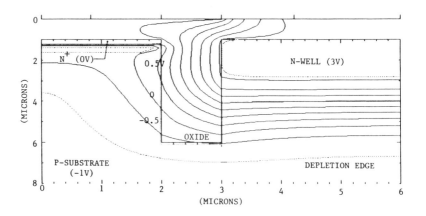

Fig. 10.11. 2-D potential profile of the trench structure, with the p-type channel implant taken into account.

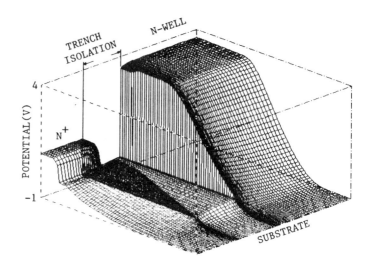

Fig. 10.12. Bird's-eye-view of the potential profile. The p-type channel implant is taken into account.

Trench Isolation

QSS (/CM2)	NA (/CM3)	VW (V)	VB (V)	INVERSION
0	6E14	3	−1	NO
5E10	6E14	3	−1	NO
1E11	6E14	3	−1	YES
1E11	2E15	3	−1	NO
3E11	5E15	3	−1	YES
3E11	1E16	3	−1	NO
1E11	6E14	3	0	YES
1E11	6E14	3	−1	YES
1E11	6E14	3	−3	YES
1E11	6E14	0	−1	NO

QSS=POSITIVE FIXED CHARGE DENSITY
NA=BULK DOPING CONCENTRATION
VW=N−WELL BIAS; VB=SUBSTRATE BIAS

Table 10.1. Summary of trench simulations.

ranges from 0 to 3E11 cm^{-2}, and the bulk doping concentration ranges from 6E14 to 1E16 cm^{-3}. N-well biases of 0 V and 3 V and substrate bias of 0 to -3 V are considered. Table 10.1 summarizes the results of the simulations. The cases to be considered are put into four groups. The first group studies the effect of increasing Q_{ss}. The second group studies the effect of increasing substrate doping. The third group studies the effect of substrate bias and the last studies the effect of n-well bias. Also indicated in the last column is a qualitative description of the simulated results. The n$^+$ region is biased at 0 V. In all cases, the p-type channel implant overlapping the n$^+$ region is taken into account.

From the first three rows of Table 10.1, the results show that with a substrate bias of -1 V, and p-substrate doping of 6E14 cm^{-3}, trench surface inversion occurs if Q_{ss} is greater than 5E10 cm^{-2}. The second group, from the third to sixth row, shows that for a Q_{ss} of 1E11 cm^{-2} and 3E11 cm^{-2}, the substrate doping has to be raised to 2E15 cm^{-3} and 1E16 cm^{-3} respectively in order to avoid trench surface inversion. The substrate bias effect is shown from the 7th to the 9th row. The substrate bias shows little effect in preventing inversion if Q_{ss} is 1E11 cm^{-2} and p-substrate doping is 6E14 cm^{-3}, although it is found that at lower Q_{ss} values, negative substrate bias reduces the probability of inversion. The last row shows that for a Q_{ss} value of 1E11 cm^{-2}, the inversion can be avoided if n-well is biased at zero volt. This merely serves to reveal the effect of the n-well bias. The n-well has to be biased to a positive voltage for the circuit to function, and the results show that minimizing the n-well bias will reduce the probability of inversion.

10.5 Experimental Results

The trench surface inversion problem has been studied experimentally and compared with the simulated results. The problem of trench surface inversion is manifested in the subthreshold characteristics of the n-channel transistor, as shown in Fig. 10.13. The transistor which is essentially adjacent to the trench shows severe leakage problem, while the transistor at a distance of 1.5 μm away from the trench has no leakage problem, and is electrically indistinguishable from LOCOS isolated transistors. The p-channel transistor has no leakage problem, and is similar to non-trench devices. This indicates that the problem is positive charge on the trench surface or in the refilled material which causes inversion along the trench surface in the p-substrate region.

The effective charge density at the trench surface was determined by the trench surface inversion test structures [10.7]. The value was found to be 1.8E11 cm^{-2}. Comparison with Table 10.1 of the simulation results shows that a p-substrate doping concentration of at least 5E15 to 1E16 cm^{-3} is necessary to avoid inversion, which means that p-well CMOS is a better choice for the application of trench isolation.

Experimental results have also shown that when the trench surface is inverted, the latchup resistance of the trench structure is not as high as expected. Table 10.2 shows the results for the measurements of the latchup

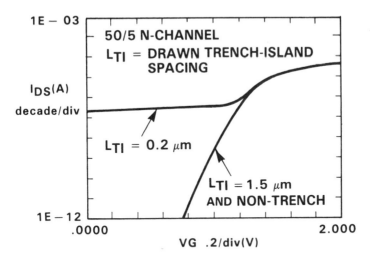

Fig. 10.13. Subthreshold behavior of trench isolated n-channel MOSFET, with different trench to island spacing. Non-trench device data is also shown for comparison.

COMPARISON OF LATCH−UP DATA TRENCH VS. NON−TRENCH

(CT029−12,17)

	B(vert)	B(lat)	I(in) (mA)	I(h) (mA)
TRENCH	30	6−9	1.5−2.5	3−4
NON−TRENCH	50	4.5−7.5	0.3−0.5	2−2.5

Table 10.2. Latchup data for trench and non-trench structures. B(vert) and B(lat) are the vertical and lateral gain of the parasitic bipolar transistors. I(in) and I(h) are the latchup initiating and holding currents.

initiating and holding current levels for the trench and non-trench structures. The difference is quite small, although the trench structure has an n⁺ to p⁺ spacing of 4 μm, which is much smaller than the conventional test structure, of 8 μm. The latchup resistance of the trench structure is much less than expected in this case, and is due to the existence of the inverted channel along the trench, all the way to the n-well. If this conducting path to the n-well is disconnected, for example, by a p-type implant into the trench [10.4], the latchup resistance will be recovered and would be much better than the conventional LOCOS structures.

10.6 Summary

By using two-dimensional simulations, the problem of trench surface inversion has been studied in detail to investigate the possibility of placing the transistors adjacent to the trench. Different conditions of charge density, biases, and doping concentrations have been considered. The regions of inversion have been identified. The results show that if Q_{ss} is larger than 5E10 cm⁻², then inversion will likely occur, unless the p-substrate doping concentration is increased. If Q_{ss} is larger than 1E11 cm⁻², then p-well CMOS is more suitable for the application of trench isolation, due to the much higher doping concentration in the p-well. By minimizing the n-well bias, the probability of inversion is also reduced. Experimental results indicated Q_{ss} values of about 2E11 cm⁻². Latchup resistance of the trench isolated structures with the transistors adjacent to the trench side-wall showed only slight improvement over the LOCOS isolated structures, under the condition of surface inversion. This indicates that the inversion path between the n⁺ region and the n-well has to be disconnected in order to recover the trench resistance to latchup. This can be done either by implanting into the trench, or by using p/p⁺ epi where the trench bottom is within the p⁺ region. Trench isolation is more suitable for p-well CMOS, where the inversion problem can be put under control if Q_{ss} is not too large. If body effect and diffusion capacitance are not of major concern, then trench isolation can also be used for n-well CMOS or NMOS processes by using a lower resistivity substrate, or using high dose and deep boron channel implants. If inversion is a major problem, then the n-channel transistor has to be put at a distance of about 1 μm away from the trench, where experimental results showed that leakage is not a problem.

References

[10.1] S.-Y. Chiang, K. M. Cham, D. W. Wenocur, A. Hui, and R. D. Rung, "Trench Isolation Technology for MOS Applications," *Proc. of the First International Symposium on VLSI Science and Technology*, 1982, pp. 339-346.

[10.2] R. D. Rung, H. Momose, and Y. Nagakubo, "Deep Trench Isolated CMOS Devices," *Tech. Digest of IEDM 1982*, pp. 237-240.

[10.3] K. M. Cham and S.-Y. Chiang, "A Study of the Trench Surface Inversion Problem for the Trench CMOS Technology," *IEEE Electron Device Lett.*, **EDL-4**, Sept 1983, pp. 303-305.

[10.4] T. Yamaguchi, S. Morimoto, G. H. Kawamoto, H. K. Park, and G. C. Eiden, "High Speed Latchup-Free, 0.5 μm-Channel CMOS Using Self-Aligned TiSi$_2$ and Deep Trench Isolation," *Tech. Digest of IEDM 1983*, pp. 522-525.

[10.5] D. B. Estreich, "The Physics and Modeling of Latch-Up in CMOS Integrated Circuits," Stanford Electronics Labs Report #G-201-9, 1980.

[10.6] R. R. Troutman, "Recent Developments in CMOS Latchup," *Tech. Digest of IEDM 1984*, pp. 296-299.

[10.7] K. M. Cham, S.-Y. Chiang, D. W. Wenocur, and R. D. Rung, "Characterization and Modeling of the Trench Surface Inversion Problem for the Trench Isolated CMOS Technology," *Tech. Digest of IEDM 1983*, pp. 23-26

[10.8] C. W. Teng, C. Slawinski and W. R. Hunter, "Defect Generation in Trench Isolation," *Tech. Digest of IEDM 1984*, pp. 586-589.

[10.9] S. M. Sze, *Physics of Semiconductor Devices*, second ed., New York:Wiley-Interscience, 1981.

[10.10] A. S. Grove, *Physics and Technology of Semiconductor Devices*, New York:Wiley, 1967.

Chapter 11

Development of Isolation Structures for Applications in VLSI

11.1 Introduction to Isolation Structures

Device isolation has become one of the major issues in VLSI. As more and more devices are packed together on a single chip, the spacing between devices is reduced significantly. Island width/space design rules are becoming very aggressive, in the range of 1 μm/1 μm [11.1]. This means that the width of the isolation structures has to be scalable without causing field leakage problems. Also, as the transistor widths are scaled down, to the range of 2 μm or less, narrow width effects become a major issue [11.2]-[11.5]. These effects are dependent on the isolation structures since the channel width of the device is defined by the field isolation. Many novel isolation structures have been investigated for applications in VLSI [11.6]-[11.14].

In general, there are several considerations in the design of isolation structures. First, the isolation dielectric thickness should be as thick as possible. This will provide a higher field threshold voltage, as well as a lower parasitic capacitance for conducting lines over the field oxide. Second, the field isolation should not encroach significantly into the island area (active area). In VLSI, over a million transistors may be packed on a chip, which

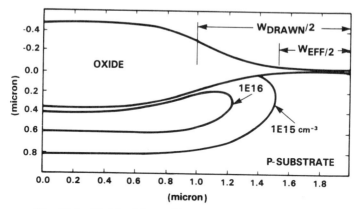

Fig. 11.1. Field oxide and boron encroachment in LOCOS.

means that the transistor size is a major consideration. Drawn channel widths (i.e., mask dimensions) are being scaled to the 1 μm range in advanced integrated circuit development [10.15]. Any encroachment of the field isolation into the active area will cause a significant reduction in the current drive of the device. This problem is shown schematically in Fig. 11.1 for the case of LOCOS isolation. [11.16]. The effective channel width of the device is significantly smaller than the drawn width. The slope of the field oxide at the boundary of the active area is also a consideration. If the slope is shallow as is the case for LOCOS, then this causes an increase in the threshold voltage of the transistor as the device width is scaled down. The threshold increase occurs because the gate electric field spreads into the field area, causing an increase of the effective channel doping of the device. Also, if a lightly doped substrate is used, a field implant of an impurity of the same type as the substrate is necessary to produce the desired field threshold voltage. In this case, the impurity may diffuse into the active area during processing, thus causing an increase in the threshold voltage or a loss of effective channel width of the active transistor. These problems of oxide and boron encroachments will be discussed in more detail later for the case of LOCOS isolation. Third, the isolation structure is preferred to be planar with respect to the silicon surface so that subsequent steps in photolithography and etching will not be affected by topological problems. A planar surface, for example, will minimize necking of

Isolation Structures

narrow polysilicon lines running across an island/field edge. The last consideration, but not the least, is process complexity. It is highly desirable that any novel isolation structures be compatible with existing MOS processes, and not require additional masking steps.

From the above discussions, it can be seen that in most cases, compromise must be made to optimize the design of the isolation structure. Simple isolation techniques such as LOCOS result in oxide encroachment. Higher implant dose to maintain the field threshold results in higher diffusion capacitance. The use of thicker field oxide to reduce interconnect capacitance may result in more oxide encroachment as well as more complicated processing problems due to topology. The following sections describe the design of isolation structures, using simulation programs as tools to optimize the design. The discussion begins with the simplest isolation technique, LOCOS. Then, modifications of this technique to minimize oxide and boron encroachment are discussed. Finally, a new isolation technique, SWAMI, is presented, which will be shown to solve most of the concerns mentioned above.

11.2 Local Oxidation of Silicon (LOCOS)

LOCOS [11.16] has been used in most integrated circuits up till this date. The process is very simple and well known. Only a brief description will be presented. The process is shown in Fig. 11.2, using the output of the SUPRA simulation. A layer of thin oxide (stress relief oxide) is grown above the silicon substrate. This is followed by a layer of nitride. Then the nitride is patterned using photolithography technique, followed by plasma etching. This defines the regions of field and active area. Then a field implant is performed to increase the threshold voltage at the field. The photoresist is then removed followed by oxidation of the exposed silicon area. The simulation shows that the encroachment of the oxide is very significant. For this case, the encroachment is about 0.3 μm on each side for a field oxide thickness of 0.55 μm. This oxide encroachment is mainly due to the presence of the stress relief oxide underneath the nitride, which is necessary to avoid defects in the active area of the silicon. The oxygen diffuses through this oxide layer during field oxidation, and hence oxide

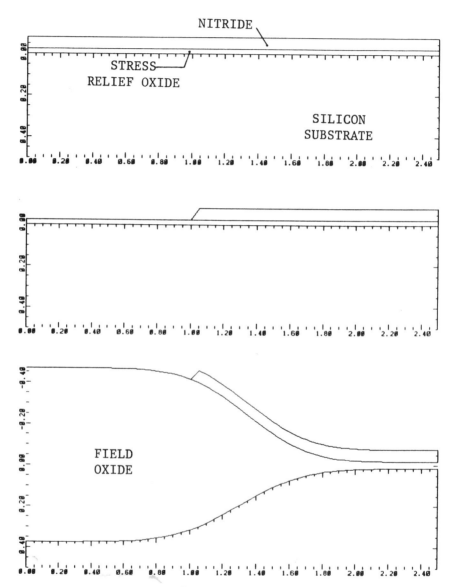

Fig. 11.2. LOCOS process simulated by SUPRA.

Isolation Structures

is grown under the nitride layer near the field edge.

Another concern with LOCOS is the encroachment of the boron impurities into the active area. If the impurity concentration underneath the active area is significantly changed, then this will cause a reduction of effective channel width, and may also cause an increase in the threshold voltage of the active transistor. In order to see how serious is the problem, the SUPRA program is used to simulate the 2-D structure, and the output data is coupled to the GEMINI program to calculate the current-voltage characteristics of the transistor in the linear region. Fig. 11.3 shows the simulated cross-section of the width of the transistor. The LOCOS isolation structure together with the field and channel implant impurity profile are shown. The simulated depletion region indicates that part of the gate electric field has terminated into the field region, hence there will be an effect on the electrical characteristics of the device.

Fig. 11.4 shows the simulated current-voltage characteristics of transistors with varying drawn widths (W_D). I_{DS}/W_D is plotted versus V_{GS} for drawn channel widths from 1 to 4 μm as well as for a very wide device of 104 μm. If the isolation is ideal, such that the electrical channel width and the drawn channel width are the same, then all of the curves will be identical to that of the wide device, since I_{DS} is proportional to the effective channel width. The simulated results show that for LOCOS isolation, this is not the case. The value I_{DS}/W_D decreases rapidly as W_D is reduced to 1 μm. This clearly shows that LOCOS is not acceptable for VLSI.

The effective width of the transistors can be calculated from these simulated results. Since the transconductance, given by the slope of the I_{DS} vs. V_{GS} curve, is proportional to the effective channel width of the device, a plot of the transconductance versus the drawn channel width will provide us the value of the channel width loss due to the isolation process. Fig. 11.5 shows the extraction of the channel width. The actual width is found to be 0.7 μm less than the mask width.

The channel width loss can be due to both the oxide encroachment and the boron field implant which diffuses into the active area, causing an increase of the threshold voltage at the active area edge. The latter effect can be studied by comparing the simulated current-voltage characteristics of a narrow

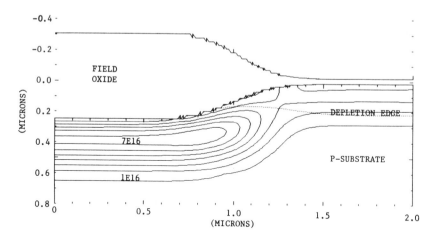

Fig. 11.3. LOCOS isolated channel width simulation. The boron distribution due to the field and channel implants, as well as the depletion edge are shown.

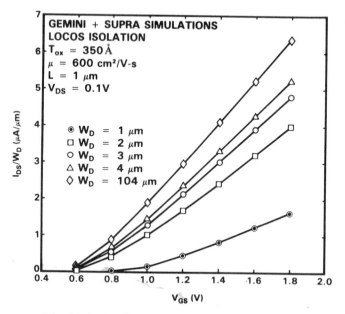

Fig. 11.4. I_{DS} / W_D vs. V_{GS} for LOCOS isolation.

Isolation Structures

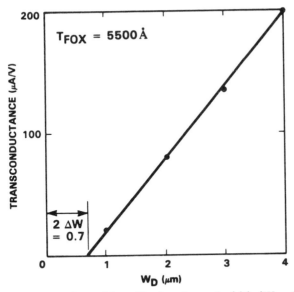

Fig. 11.5. Extraction of the effective channel width ($W_D - 2\Delta W$).

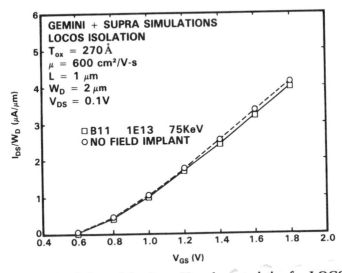

Fig. 11.6. Simulation of the $I_{DS} - V_{GS}$ characteristics for LOCOS isolated MOSFET, with and without field implant.

width device with and without field implant. Fig. 11.6 shows that the field implant only reduces the current slightly, hence the device channel width loss is mainly due to the oxide encroachment. In the case where oxide etch back is used to reduce the bird's beak, the effect of boron encroachment will be more significant. This will be discussed in the next section.

Another figure of merit for an isolation technology is the sensitivity of the threshold voltage to the device width. From the circuit design point of view, it is highly desirable that the threshold voltage be insensitive to the device geometry. Fig. 11.7 shows the simulated results of the threshold voltage versus the drawn channel width for the LOCOS isolation. The threshold voltage begins to increase rapidly for drawn width below 2 μm. This is because the gate electric field partially terminates under the field oxide at the edge of the active area. This can be observed in Fig. 11.3 which shows the channel depletion edge. This again indicates that the application of LOCOS is limited to a minimum device width of at least 1.5 μm for optimum performance.

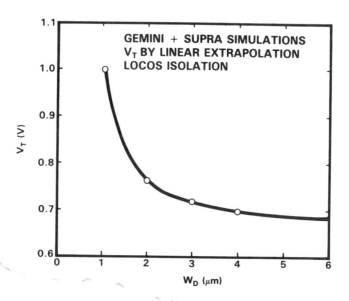

Fig. 11.7. Simulated threshold voltage vs. W_D.

11.3 Modified LOCOS

One way to reduce the oxide encroachment (bird's beak) is to grow a thick field oxide and then etch back to a thinner thickness. The idea can be understood by looking at a field oxide profile generated by a two dimensional oxidation program SUPRA, as shown in Fig. 11.2. It can be seen from the profile that the bird's beak is long but the initial slope is quite flat. Etching away 200 nm to 300 nm of field oxide can hence reduce the bird's beak significantly. In this section, an optimum design of this etch back LOCOS process using one-dimensional process simulation program SUPREM and two-dimensional simulation program SUPRA will be discussed. The goal is to minimize both the boron and oxide encroachments.

The original process uses a stress-relief oxide thickness of 40 nm, a nitride thickness of 200 nm and a boron field implant dose of 5E13 cm^{-2} to obtain a final field threshold voltage of about 15 V. 800 nm of field oxide is grown at 950 deg C in steam for 300 minutes and then etched back to 500 nm. Fig. 11.8 shows the simulated final field oxide and doping profile from SUPRA. The vertical line indicates the position of the bird's beak before etch back. The oxide encroachment is reduced by about 0.28 microns and the active area is about the same as originally defined by the nitride. However, the boron encroachment effect is quite significant. The dotted line in Fig. 11.8 shows that due to field implant the boron concentration line of 1E17 cm^{-3} is quite a distance into the active area. Simulation of the active area boron concentration due to threshold and punchthrough control implant is plotted as a function of distance into the silicon substrate in Fig. 11.9. The surface boron concentration is about 8E16. Therefore, a significant part of the active area has a higher boron concentration due to the lateral diffusion of the boron field implant. In this case, the boron encroachment actually dominates over the oxide encroachment. The effective channel width is not much improved by this etch back process.

In order to solve the boron encroachment problem the process is modified and simulated again. A thinner field oxide is grown so that both the thermal cycle and the field implant boron dose can be reduced. In the modified process, 600 nm of field oxide is grown and then etched back to about

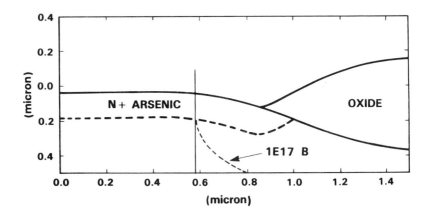

Fig. 11.8. Oxide and impurity profile after etch back.

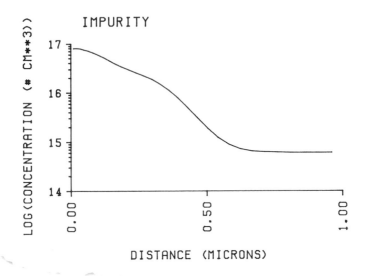

Fig. 11.9. Channel impurity profile vs. vertical distance.

Isolation Structures

500 nm. SUPREM is used to find the oxidation cycle for 600 nm of field oxide and the boron dose needed to give the same field threshold voltage of 15 V. It is found that the field implant dose can be reduced from 5E13 to 3E13 cm^{-2} and the thermal cycle can be reduced to 200 minutes at 950°C. SUPRA is then used to simulate the entire new process. The resulting field oxide and doping profile is shown in Fig. 11.10. The field oxide encroachment in this case is about 0.15 µm more than in the previous process. However, the boron encroachment is about 0.13 µm less, using the 1E17 cm^{-3} boron line as a reference. Overall, there is a gain of 0.1 µm over the previous process. Hence, in the design of an optimum LOCOS process, both the oxide and boron encroachment need to be carefully taken into account.

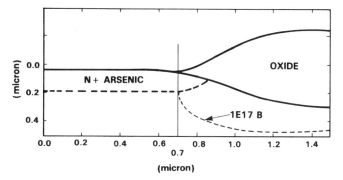

Fig. 11.10. Oxide and impurity profile for modified LOCOS.

11.4 Side Wall Masked Isolation (SWAMI)

The problems of oxide and boron encroachments associated with LOCOS must be minimized in VLSI. New isolation techniques have to be designed such that the considerations discussed in the introduction are satisfied in order to achieve optimum circuit performance. A new isolation process, SWAMI (Side Wall Masked Isolation), has been developed at Hewlett-Packard Laboratories [11.8]. This process effectively eliminates the bird's beak during field oxidation while maintaining a desirable field oxide thickness.

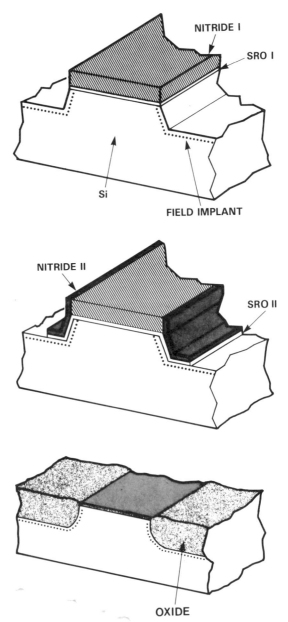

Fig. 11.11. The SWAMI process.

Isolation Structures

Fig. 11.11 shows the sequence of steps in the fabrication process used to form the SWAMI isolation structure. For more detailed descriptions, the reader is referred to the reference article. The objectives of the process are to prevent field oxide encroachment into the active area by using a second layer of thin nitride on the side wall and to reduce stress by using a sloped side wall. The sloped side wall is used to prevent defect formation due to volume expansion induced stress during field oxidation. SEM cross-sectional view of the isolation structure did not reveal any bird's beak formation in the active area. The sloped side wall is produced by C_2F_6 plasma etching [11.8], which introduces a loss of island width of about 0.1 μm per side. Another factor that might contribute to the channel width loss is the boron encroachment, since a field implant step is necessary to eliminate the problem of field inversion, especially at the sloped side wall. In this section, simulations will be used to compare the performance of narrow width devices using an ideal vertical isolation, the SWAMI, and LOCOS. The effect of boron encroachment for the SWAMI process is also simulated.

To simulate the SWAMI oxide structure, a program with the capability of simulating generalized isolation structures is needed. This can be done by the combination of SUPRA and SOAP. The SOAP program is used to generate the isolation structure. The output data file is then loaded into the SUPRA file for the impurity profile simulations. Fig. 11.12 shows the isolation structure generated by SOAP for the SWAMI process. The sloped silicon side wall and the nitride layer covering the side wall have significantly reduced the oxide encroachment into the active area, although there is still a bird's beak formed at the island edge. Since SEM pictures have shown experimentally that there is essentially no bird's beak at the island edge, the oxide growth near that region is overestimated by the SOAP program. One possibility is the effect of stress in that region which might have reduced the oxygen diffusivity [11.17].

The SOAP program also simulates the stress along the oxide/silicon interface as a function of time during oxidation. This is useful if one is concerned about the stress induced by the isolation structure. In the case of SWAMI, stress may be a concern at the island edge covered by the nitride, where volume expansion induced stress occurs during field oxidation. Fig. 11.13 shows the stress calculations using the SOAP program for the

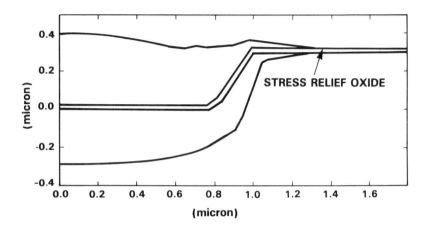

Fig. 11.12. SWAMI structure simulated by SOAP.

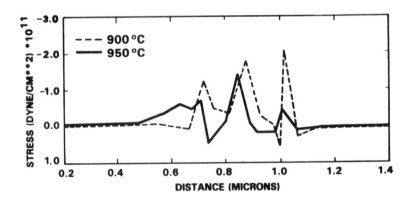

Fig. 11.13. Stress simulation for SWAMI using SOAP.

Isolation Structures

SWAMI process. Here positive stress means compression and negative stress dilation. The stress calculated is after 0.5 μm of field oxide is grown. The horizontal axis corresponds to the horizontal distance from the left edge of the structure. The results show that stress occurs along the sloped side wall covered by the thin nitride. As oxide is grown, stress is induced by volume expansion. The results also show that stress can be reduced by using higher oxide growth temperature. This is due to the lower viscosity of the oxide, hence more flow is possible to reduce the stress.

In order to see if boron encroachment is a problem in SWAMI, the impurity distribution has to be simulated. This can be done using the SUPRA program. The program accepts the data file from the SOAP program which defines the final oxide structure. The SUPRA program then calculates the impurity distribution associated with the oxide boundary conditions and the temperature cycle of the field oxidation. Then the channel implants and subsequent process steps can follow. Fig. 11.14 shows the simulated structure with the impurity contours without the channel implants. This shows the extent of the boron distribution due to field implant. From this figure, it can be seen that the boron concentration under the active area is not high, in comparison with typical channel implants, especially for submicron devices (Fig. 12.19). The effect on the threshold voltage and channel width narrowing should be only slight.

Since the SOAP program overestimates the oxide encroachment, the device performance simulated by using this structure will not be accurate. The electrical effects of the oxide shape can be approximated by the GEMINI program. If the depletion region of the device does not extend too deeply into the substrate, then it is only necessary to approximate the side wall of the oxide structure. The impurity profile of the structure can be first simulated by SOAP-SUPRA combination, then copied into the GEMINI input file by using a Gaussian profile approximation. This procedure is somewhat lengthy, but should provide better agreement with experiments.

Fig. 11.15 shows the simulation of the SWAMI structure for a MOSFET with channel width of 2 μm. Only half of the device needs to be simulated due to symmetry. The GEMINI simulation provides a good representation of the upper part of the oxide structure, which determines most of the narrow width

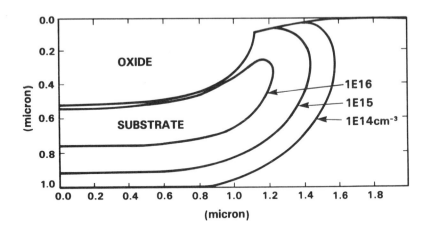

Fig. 11.14. Impurity profile simulation for SWAMI.

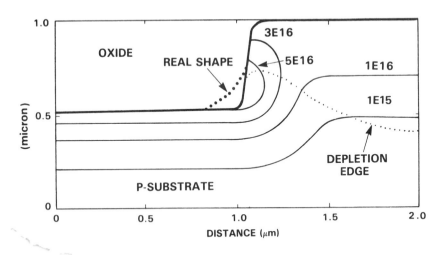

Fig. 11.15. Channel width simulation of SWAMI by GEMINI.

Isolation Structures

effects. Note that the depletion region ends near the top of the isolation, hence any approximation of the oxide shape near the bottom will not give rise to significant errors. The drain to source current as a function of the gate bias can be simulated. From the current-voltage characteristics, the threshold voltage as well as the transconductance can be calculated. This provides information about the narrow width effects, which are manifested as increase in threshold voltage and reduction in transconductance.

This procedure is repeated for the cases of an ideal vertical isolation, and also for LOCOS. The effect of boron encroachment can be studied by simulating structures with and without field implants. Oxide thickness of 500 nm is considered. Fig. 11.16 shows the simulated current voltage characteristics of the different cases. I_{DS} is plotted versus V_{GS} for the devices biased in the linear region. From the slopes of the curves, the effective width of the devices can be calculated. SWAMI is slightly inferior to the ideal vertical isolation, but much better than LOCOS. Note that the main effect of the boron field implant on the SWAMI structure is to increase the threshold voltage, while the transconductance is not significantly changed. For the case of LOCOS, the reduction in current is mostly due to the oxide encroachment into the active area. The effect of boron encroachment is insignificant. For the SWAMI process, the channel width loss is almost independent of the field oxide thickness up to about 0.7 μm, while the LOCOS has a width loss per side approximately 70% of the oxide thickness. Hence for thicker field oxide, which is desirable for reducing the interconnection parasitic capacitance, isolation structures such as SWAMI are absolutely necessary. Table 11.1 summarizes the results of the simulations. The channel width loss for the SWAMI process is found to be 0.11 μm per side due to oxide encroachment, and about 0.02 μm per side due to boron encroachment.

11.5 Summary

Simulations have been found to be very useful in developing and understanding isolation structures. The electrical characteristics of narrow width devices can be understood and optimized by coupling process simulations and

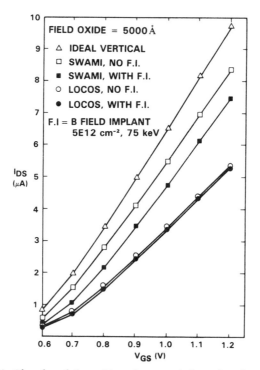

Fig. 11.16. Simulated I_{DS} - V_{GS} characteristics of n-channel MOSFET with drawn width of 2 μm, for vertical, SWAMI and LOCOS isolations.

ISOLATION STRUCTURE	DW (MICRONS) PER SIDE
IDEAL VERTICAL	0
LOCOS, NO F.I.	0.35
LOCOS, WITH F.I.	0.36
SWAMI, NO F.I.	0.11
SWAMI, WITH F.I.	0.13

F.I.= BORON FIELD IMPLANT
FIELD OXIDE THICKNESS = 5000A

Table 11.1. Summary of MOSFET width simulations.

device simulations. Advanced isolation structures such as SWAMI can be simulated; SWAMI's effect on device characteristics has been studied and compared with other isolation structures. As isolation structures evolve to higher complexity, the use of numerical simulations is essential for efficient process and device development.

References

[11.1] O. Kudoh, H. Ooka, I. Sakai, M. Saitoh, J. Ozaki and M. Kikuchi, "A New Full CMOS SRAM Cell Structure," *Tech. Digest of IEDM 1984*, pp. 67-70.

[11.2] T. Yamaguchi and S. Morimoto, "Analytical Model and Characterization of Small Geometry MOSFET's," *IEEE Trans. on Electron Devices*, **ED-30**, June 1983, pp. 559-571.

[11.3] L. A. Akers, M. M. E. Beguwala, and F. Z. Custode, "A Model of a Narrow-Width MOSFET Including Tapered Oxide and Doping Encroachment," *IEEE Trans. Electron Devices*, **ED-28**, Dec 1981, pp. 1490-1495.

[11.4] C.-R. Ji and C.-T. Sah, "Analysis of the Narrow Gate Effect in Submicrometer MOSFET's," *IEEE Trans. Electron Devices*, **ED-30**, Dec 1983, pp. 1672-1677.

[11.5] F. H. Gaensslen, "Geometry Effects of Small MOSFET Devices," *IBM J. Res. Develop.*, **23**, Nov 1979, pp. 682-688.

[11.6] R. D. Rung, H. Momose, and Y. Nagakubo, "Deep Trench Isolated CMOS Devices," *Tech. Digest of IEDM 1982*, pp. 237-240.

[11.7] K. L. Wang, S. A. Saller, W. R. Hunter, P. K. Chatterjee, P. Yang, "Direct Moat Isolation for VLSI," *IEEE Trans. Electron Devices*, **ED-29**, Apr 1982, pp. 541-547.

[11.8] K. Y. Chiu, J. L. Moll, K. M. Cham, J. Lin, C. Lage, S. Angelos, and R. C. Tillman, "The Sloped-Wall SWAMI--A Defect-Free Zero Bird's Beak Local Oxidation Process for Scaled VLSI Technology," *IEEE Trans. Electron Devices*, **ED-30**, Nov 1983, pp. 1506-1511.

[11.9] N. Matsukawa, H. Nozawa, J. Matsunaga, S. Kohyama, "Selective Polysilicon Oxidation Technology for VLSI Isolation," *IEEE Trans. Electron Devices*, **ED-29**, Apr 1982, pp. 561-567.

[11.10] S. Hine, T. Hirao, S. Kayano, N. Tsubouchi, "A New Isolation Technology for Bipolar Devices by Low Pressure Selective Silicon Epitaxy," *Tech. Digest of Symposium on VLSI Technology 1982*, pp. 116-117.

[11.11] N. Endo, K. Tanno, A. Ishitani, I. Kurogi, H. Tsuya, "Novel Device Isolation Technology with Selective Epitaxial Growth," *Tech. Digest of IEDM 1982*, pp. 241-244.

[11.12] T. Shibata, R. Nakayama, K. Kurosawa, S. Onga, M. Konaka, and H. Iizuka, "A Simplified Box (Buried Oxide) Isolation Technology for Megabit Dynamic Memories," *Tech. Digest of IEDM 1983*, pp. 27-30.

[11.13] S.-Y. chiang, K. M. Cham, D. W. Wenocur, A. Hui, and R. D. Rung, "Trench Isolation Technology for MOS Applications," *Proc. of the First International Symposium on VLSI Science and Technology 1982*, pp. 339-346.

[11.14] H.-J. Voss and H. Kurten, "Device Isolation Technology by Selective Low-Pressure Silicon Epitaxy," *Tech. Digest of IEDM 1983*, pp. 35-37.

[11.15] K. Nakamura, M. Yanagisawa, Y. Nio, K. Okamura, and M. Kikuchi, "Buried Isolation Capacitor (BIC) Cell for Megabit CMOS Dynamic RAM," *Tech. Digest of IEDM 1984*, pp. 236-239.

[11.16] J. A. Appels, E. Kooi, M. M. Paffen, J. J. H. Schatorje, and W. H. C. G. Verkuylen, *Phillips Res. Rep.*, **25**, 1970, pp. 118.

[11.17] Daeje Chin, private communication.

Chapter 12

Transistor Design for Submicron CMOS Technology

In this chapter, the design of transistors for submicron CMOS technology will be presented. The advantages of, as well as issues involved in CMOS technology will first be discussed. Then the concerns for the design of n- and p-channel MOSFET's with submicron channel lengths will be discussed. Using simulations, the values of the critical device parameters are determined which will minimize leakage problems in submicron transistors.

12.1 Introduction to Submicron CMOS Technology

Before getting into the details of process development for CMOS circuits using CAD tools, it is useful to understand why we should put so much effort into CMOS. CMOS technology has gained more and more significance, overtaking NMOS technology as the era of VLSI approaches. There are several major reasons. First, when the number of transistors on a chip approaches a million, the power consumption becomes a limiting factor. For NMOS technology, creative circuit design is necessary to minimize power consumption, and expensive packaging is needed for heat dissipation. CMOS circuits, due to the complementary nature of the n- and p-channel transistors, consume much less power, especially in the standby state, where the power consumption is determined by the leakage current of the transistors biased in the off state. The power consumption of the circuit in the active state depends on the switching frequency of the circuits. For example, in the case of 16K static RAM, using an access time of 70 ns as the point of comparison, the CMOS chip dissipates approximately 130 mW while the NMOS chip dissipates 280 mW [12.1]. Low power operation means simplicity and low cost in packaging, both at chip level and system level. In a battery operation environment, the use of CMOS circuits is essential for minimum power consumption.

In the past, NMOS technology has enjoyed a simpler process as compared to CMOS. In a typical CMOS process, additional masks for the well, field, channel, and source/drain implants are necessary to fabricate both the n and p-channel transistors in the same circuit. But as the devices are scaled down in dimension, the device structures become much more complicated, and the process for NMOS becomes significantly more complex. Two channel implants instead of one, side-wall spacers, tip-implants in addition to the conventional source/drain implants, two-level metallization, and more advanced isolation techniques, are being employed in VLSI circuits. It becomes clear that for NMOS, the advantage of process simplicity over CMOS has been reduced significantly. The additional processing steps used in CMOS circuits are now only a small percentage of the total processing steps.

In the CMOS circuit, the p-channel transistor has lower performance compared with the n-channel transistor because of the lower mobility of holes (current carrier for p-channel) than of the electrons (current carrier for n-channel), with 200 versus 600 cm^2/V-sec. In the long channel device, the difference in current drive is about a factor of three. Thus CMOS circuits are traditionally slower than NMOS circuits. In order to have symmetry in the inverter transfer characteristics, it is necessary to make the p-channel transistor much wider than the n-channel to compensate for the mobility difference. This means a disadvantage in packing density compared with NMOS circuits. But as the channel lengths of the transistors are scaled down, this factor has become less important due to velocity saturation effects of the n-channel device.

As has been shown in Chapter 7, the current-voltage characteristics of long channel MOSFET's for drain bias below saturation can be approximated by the simple equation:

$$I_{DS} = \frac{W}{L}\mu C_{ox}[(V_{GS} - V_T)V_{DS} - \frac{1}{2}V_{DS}^2] \qquad (12.1)$$

It can be seen that the current drive is proportional to the mobility, and that the transconductance (g_m) in the saturation region is given by (see Eq. 7.4):

$$g_m \equiv \frac{\partial I_{DS}}{\partial V_{GS}} = \frac{W}{L}\mu C_{ox}(V_{GS} - V_T) \qquad (12.2)$$

We see that g_m is proportional to the mobility multiplied by the factor (V_{GS}-V_T) when the device is in the saturation regime, and is inversely proportional to the channel length. But when velocity saturation occurs along the entire channel, the current-voltage characteristics are given by [12.2]:

$$I_{DSAT} = W C_{ox} V_s (V_{GS} - V_T) \qquad (12.3)$$

where I_{DS} is proportional to the saturation velocity (V_s). The corresponding g_m is a constant, and is independent of the channel length. Due to the higher mobility of the electrons, velocity saturation occurs at a lower electric field, therefore at longer channel length than for the holes, at the same drain bias. This means that as the channel lengths of n- and p-channel transistors are scaled down, the current drive of the n-channel device tends to saturate to a constant value independent of the channel length, while the p-channel device will gain in current drive until its current carriers exhibit the velocity saturation effect, at a shorter channel length. The comparison of n and p-channel transistor current drives measured as a function of L_{eff} is shown in Fig. 12.1. The current drive of the n-channel transistor begins to deviate from the $1/L_{eff}$ dependence due to velocity saturation, while the p-channel transistor current drive still obeys the $1/L_{eff}$ dependence. Because of this effect, the difference between n- and p-channel performance is reduced as the devices are scaled down.

The density and speed of CMOS circuits can also be enhanced by creative circuit designs such as the domino-CMOS [12.3], where only a small percentage of the devices are p-channel. In this case, the speed and density of

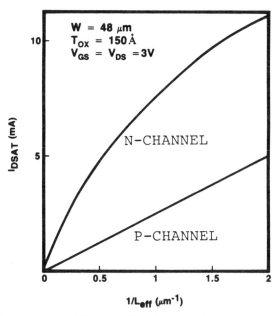

Fig. 12.1. Current drive vs. $1/L_{eff}$ for n and p-channel MOSFET's.

the circuit is very close to NMOS circuits, while the power consumption is much reduced.

CMOS also has another major advantage, which is a very large noise margin. Unless major malfunction of the transistors has occurred, the circuit will function at least logically, although the power consumption and speed specification may be violated.

So far the advantages of CMOS circuits have been presented. However, there are also some major concerns in CMOS technology. The best known problem is latchup [12.4]-[12.5]. This subject is covered in more detail in Chapter 10, which discusses the use of trench isolation to prevent latchup, and the problems that are associated with this technique. The other concerns of submicron CMOS technology are device issues, which are of a more general nature.

The design of short channel transistors is complicated by the effect of drain-induced barrier lowering (DIBL), as described in Ch. 8. Although the physics of DIBL in n-channel transistors has been discussed in that chapter, the

DIBL that occurs in the p-channel transistor is somewhat more complicated. This will be discussed in detail in the next section. Other device concerns such as the drain to source avalanche breakdown of the n-channel transistors [12.6] (the p-channel transistor does not have this problem as severely as the n-channel because of the much lower impact ionization coefficient of the holes), as well as device degradation due to hot carrier effects [12.7] are of a general nature, not specific to CMOS. In Chapter 9, the simulation of device structures used to reduce the hot carrier effects has been presented. In this chapter, the major emphasis will be on minimizing the residual leakage current of the p-channel transistor due to DIBL.

As will be shown later, the reason for the more severe DIBL problem in the p-channel transistor is because n^+ polysilicon is used for gate material. If p^+ polysilicon is used for the p-channel transistor, then the DIBL problem for the p-channel will be reduced to the same degree as for the n-channel transistor. But a combination of the n^+ and p^+ polysilicon will cause process complications. One of them is the processing of the p^+ polysilicon gate. Boron is used to dope the polysilicon to form the p^+ gate. But boron may diffuse through the gate oxide, causing threshold shift. Since the gate oxide thickness is being scaled down, typically to 15 nm in submicron devices, this problem will become even more serious in VLSI. Another problem is the connection of the n^+ and p^+ polysilicons within the chip. The obvious choice would be the use of silicides. This would require the development of a silicide process. Also, the n^+ and p^+ dopant may diffuse through the silicide grain boundaries to compensate each other. In this chapter, n^+ polysilicon gate is assumed for the process. It will be shown that by carefully choosing the device parameters, the leakage current can be minimized at effective channel lengths of 0.5 μm. For further reduction of the channel length, p^+ polysilicon or some other gate material having the appropriate work function must to be used [12.8],[12.9].

12.2 Development of the Submicron P-Channel MOSFET Using Simulations

In this section, the development of the p-channel MOSFET will be presented. Extensive use has been made of simulations to fully understand the physics of the device. The relationship of the device structure to the DIBL effect, or the leakage current, is discussed. A guideline for the p-channel transistor structure which should reduce the leakage current will be presented.

P-Channel Transistor Structure

The p-channel transistor is fabricated on a n-type substrate or in an n-well in a p-type substrate. The fabrication procedure is similar to the n-channel transistor. The major difference in structure between the n- and p-channel is that the n-channel has a channel implant of the same impurity type as the substrate, while the p-channel implant is of the opposite type to the substrate. To understand better why this is necessary, let us consider the threshold voltage of an n-channel transistor with long channel length.

For an n^+ polysilicon gate over p-type silicon, assuming no interface charge, zero substrate bias, and a uniform bulk impurity concentration N_B, the threshold voltage is given by:

$$V_T = \phi_{ms} + 2\psi_B + \frac{\sqrt{2\varepsilon q N_B (2\psi_B)}}{C_{ox}} \qquad (12.4)$$

where the first term is the work function difference between the n^+ polysilicon gate and the p-type silicon substrate, ψ_B the potential difference between the intrinsic and hole Fermi level (≈ 0.35 V), and the last term the voltage drop across the depletion region. The value for the work function difference is approximately -0.9 V and the term due to the depletion layer charge is about 0.1 V for $N_B = 1E15$ cm^{-3} and gate oxide thickness of 15 nm. Therefore, for a reasonable threshold voltage of typically 0.7 V, the surface doping concentration must be much higher than the typical substrate used in fabrication. The reason that we want a low bulk doping concentration is to reduce the substrate bias effect and parasitic capacitance between the n^+ source/drain regions and the p-substrate.

For the p-channel transistor, the situation is different. In n-well CMOS, which will be assumed in this study, the n-well doping concentration is in the order of 1E16 cm^{-3} near the surface. This is necessary because one needs about an order of magnitude difference in impurity concentration between the n-well and the p-substrate, for process control reasons. Secondly, the n-well must have a high enough impurity concentration to avoid bulk punchthrough of the p-channel transistor, as well as to reduce latchup susceptibility by lowering the n-well resistivity (see Ch. 10). Substituting the n-well doping concentration value into the equation, the values of the three terms become -0.2, -0.7 and -0.35 respectively. It can be observed that the threshold voltage is about -1.2 V. Since a high threshold voltage means low current drive, the magnitude of this threshold voltage is too high for most applications.

In order to reduce the magnitude of the threshold voltage, a so called counter-doping technique is employed, which is an implantation of boron into the channel [12.10]. The dose is typically low, on the order of 5E11 cm^{-2}, and the profile is very shallow, typically 0.15 μm. The boron impurities are all depleted, and effectively act like a layer of interface charge. Using this simplified argument, the threshold voltage as a function of the substrate doping and counter-doping can be estimated for long channel lengths. For comparison, Fig. 12.2 and 12.3 show calculations for the threshold voltages of both n and p-channel transistors with n$^+$ polysilicon gate. The n-channel calculations provide an idea of the dependence of the threshold voltage on the impurity concentration at the surface for two different oxide thicknesses. The p-channel calculations in Fig. 12.3 show the counter-doping dose (within the silicon substrate) that is necessary to provide a threshold voltage of -0.6 V, for different n-well surface doping concentrations. This is a good example which shows that often one can use very simple models to provide quick estimates of the parameter values, at the same time gaining some physical insight into the problem.

To obtain a better understanding of the channel profile, the SUPREM program is used to simulate the channel profiles. Fig. 12.4 shows the simulated channel impurity profile of the p-channel transistor, versus distance from the silicon surface into the substrate. The counter-doping is performed by ion

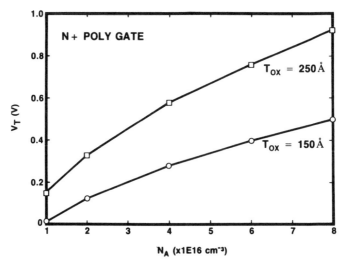

Fig. 12.2. Calculation of the n-channel threshold voltage vs. p-type substrate doping concentration for two oxide thicknesses.

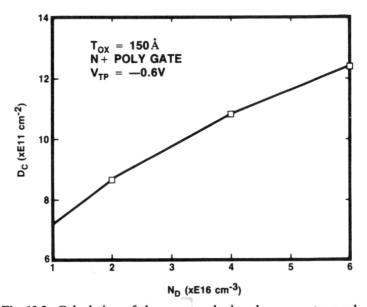

Fig. 12.3. Calculation of the counter-doping dose vs. n-type substrate concentration for a fixed p-channel threshold voltage of -0.6 V.

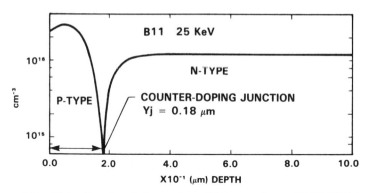

Fig. 12.4. Simulated channel profile for p-channel MOSFET.

implantation of boron (B_{11}) at 25 KeV, with a dose of 7E11 cm^{-2}. Subsequent temperature cycles during the device fabrication give rise to a junction depth of about 0.18 μm. The depth of the junction is also related to the fact that when B_{11} is implanted into silicon, there is channeling which produces a long tail on the distribution going into the substrate [12.11]. If BF_2 is used instead, then for similar implant energies, the boron atoms will have a smaller implant range, due to the heavier molecular weight. A shallower counter-doping junction is formed. This will be discussed later. Note that this implant profile is the exact opposite of what is in the n-channel transistor, as is shown in Chapter 7, Fig. 7.8. Because of this counter-doping, the effective substrate doping of the transistor can be thought of as being reduced, thus the drain electric field penetrates more towards the source, thereby giving rise to a higher susceptibility to DIBL.

Using the profile generated from the SUPREM program, the p-channel transistor structure can now be studied using the GEMINI program which solves for the potential profile within the device structure. Fig. 12.5 shows the p-channel transistor structure, with n$^+$ polysilicon gate, generated by the GEMINI program. On the left is the p$^+$ source and on the right is the p$^+$ drain. The counter-doping junction is indicated in the figure. The gate oxide thickness is 25 nm. For simplicity, the substrate doping concentration is assumed to be a constant and equal to 1E16 cm^{-3}. This is not exactly true in the n-well CMOS process, since the profile is really a Gaussian profile. But the correction to the device characteristics is small, since the device

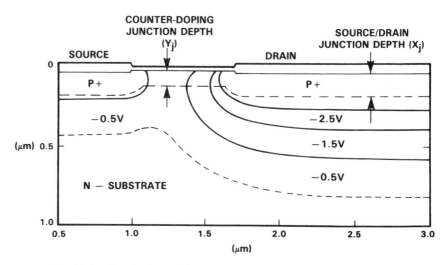

Fig. 12.5. Simulation of the p-channel MOSFET by GEMINI.

characteristics are mainly dependent on the surface concentration, down to about 0.5 μm. The bias voltage on the gate, source, drain, and substrate are 0, 0, -3 and 0 V respectively. The two-dimensional potential contours are generated by the program. The program assigns the hole potential in the substrate equal to the substrate bias (0 V here). Therefore the potentials at the p^+ source and drain regions have values of (V - 0.7 V), where V is the applied potential.

Drain-Induced Barrier Lowering in P-channel Transistors

This section describes the study of the DIBL problem in p-channel transistors with n^+ polysilicon gate, using simulations. The technique of simulating the p-channel transistor structure has been presented in the last section. Here the details of the potential distribution within the device structure are described.

The DIBL effect is dependent on three critical device parameters. They are: the gate oxide thickness (T_{ox}); counter-doping junction depth (Y_j); and source/drain junction depth (X_j), for a fixed channel length. In this discussion, the effective channel length (L_{eff}) is fixed at 0.5μm. The DIBL effect is characterized by the residual leakage current (I_L), which is the current at a drain bias of -3 V and gate bias at 0 V. Another parameter is the subthreshold slope (S), which degrades with increasing DIBL. The DIBL effect can also be characterized by the sensitivity of the threshold voltage on the channel length, when the drain is biased at -3V. The leakage current, subthreshold slope, and threshold variations are simulated as functions of the device parameters. The L_{eff} is 0.5 μm and T_{ox} is 25 nm for all simulations unless stated otherwise.

P-channel transistors with different values of Y_j (0.1, 0.15, 0.2 μm) but fixed values of threshold voltage of about -0.65 V are simulated. Since the DIBL effect is very sensitive to the threshold voltage, (the higher the magnitude of the threshold voltage, the less is the DIBL effect due to lower counter-doping dose or higher n-well doping concentration) the amount of counter-doping dose has been adjusted so that the threshold voltage is similar for the three cases to give a valid comparison. The potential profiles are shown in Fig. 12.6 for the three cases, with Y_j values increasing from top to bottom. From the potential contours, it can be observed that as Y_j is increased, the potential contours spread further from the drain towards the source. This indicates more DIBL effect. The major punchthrough path can be either at the surface or in the bulk. The bulk punchthrough path can be traced out by joining the points of the potential contours which are most extreme towards the source, as shown in the last figure of Fig. 12.6. The bulk punchthrough path is within the counter-doping junction, indicating that the counter-doping junction is enhancing the DIBL effect. Note that this is different from the case of n-channel transistors where the bulk punchthrough path under a large drain bias is quite deep ($\approx$ 0.7 μm) below the surface, due to the high surface impurity concentration and lightly doped substrate (Fig. 8.6 and Fig. 8.10 in Chapter 8).

Fig. 12.7 shows the potential profile as a function of the vertical distance from the silicon surface into the bulk, at a horizontal distance of 0.2 μm away from the source. This horizontal position is chosen because it is near the "saddle point" (see Ch. 8) of the potential distribution, hence determining the punchthrough current path. The potential (and hence also the potential barrier) in the bulk is lower for higher values of Y_j.

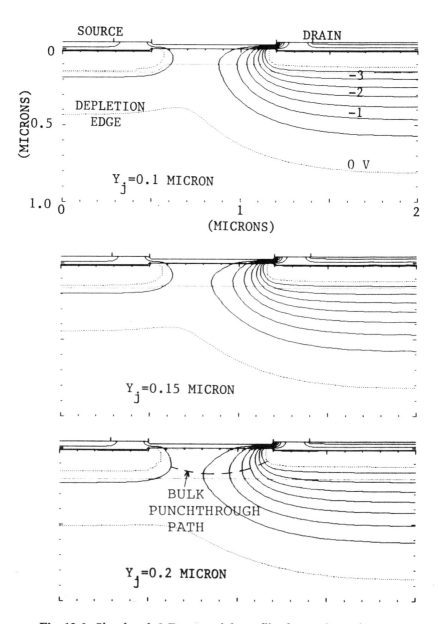

Fig. 12.6. Simulated 2-D potential profile for p-channel MOSFET, with Y_j = 0.1, 0.15 and 0.2 μm. The drain, gate bias is -3 and 0 V respectively.

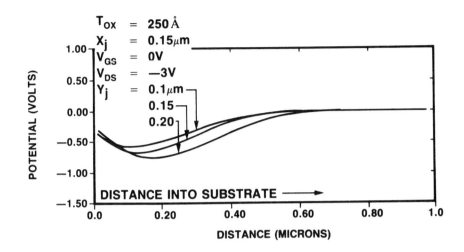

Fig. 12.7. Channel potential profile vs. vertical distance for a p-channel MOSFET with L_{eff} of 0.5μm, with three different Y_j values.

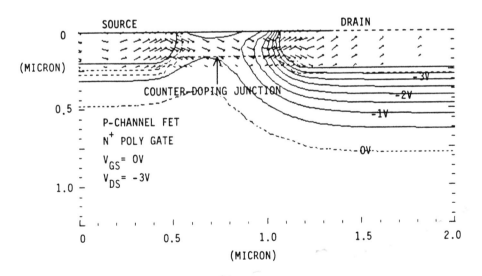

Fig. 12.8. Simulation of the punchthrough characteristics of p-channel MOSFET, using the PISCES program.

To further illustrate the punchthrough characteristics of the counter-doped p-channel transistor, the SUPREM-PISCES combination is used to simulate the current path when the device is biased at $V_{GS} = 0$ V and $V_{DS} = -3$ V and $Y_j = 0.15\,\mu\text{m}$. Fig. 12.8 shows the subthreshold leakage current, as indicated by the vectors, flowing through the counter-doped channel region. Therefore the punchthrough characteristic is very different from the n-channel transistor (see Fig. 8.10, Ch. 8), and depends on the profile of the counter-doped layer.

The subthreshold slope of the p-channel transistor under a drain bias of -3 V is simulated for different values of X_j and Y_j, and is shown in Fig. 12.9. Each curve corresponds to a fixed value of Y_j. The threshold voltage, defined by a current 1E-7($W_{\it eff}/L_{\it eff}$), is specified at each data point. The simulated data show that the subthreshold slope is strongly dependent upon X_j and Y_j.

The interpretation of the graph is as follows. With a value of $X_j = 0.1\,\mu\text{m}$, devices with different values of Y_j (0.1, 0.15, 0.2 μm) are simulated, and the counter-doping doses are adjusted to give approximately the same threshold voltage (V_T). The data show that S is very sensitive to Y_j, and increases with Y_j. The values of X_j is then increased to see the sensitivity of the device performance on X_j. It can be seen that as X_j increases from 0.1 to 0.25 μm, the magnitude of the threshold voltage decreases and S increases, both due to DIBL. The sensitivity of the device threshold voltage and subthreshold slope to X_j increases with increasing value of Y_j, indicating that the counter-doping is enhancing the short channel effect. For example, for $Y_j = 0.1\,\mu\text{m}$, an increase in X_j from 0.1 to 0.25 μm causes the subthreshold slope to increase from 104 to 122 mV/dec, while for $Y_j = 0.2\,\mu\text{m}$, the slope increases from 140 to 194 mV/dec. For a fixed value of $X_j = 0.15\,\mu\text{m}$, $V_T = -0.67$ V, and $T_{ox} = 25$ nm and 15 nm, the dependence of S on Y_j is shown in Fig. 12.10. For $T_{ox} = 25$ nm, S increases from 112 to 170 mV/dec, when Y_j increases from 0.1 to 0.2 μm, showing the effect of Y_j on the DIBL. The subthreshold slope improves for a thinner gate oxide of 15 nm, but the dependence of S on Y_j is only slightly reduced. The figure shows that the gate oxide thickness has a major effect on the subthreshold slope, as expected. It can be seen here that for submicron CMOS, thin gate oxide is essential for controlling the subthreshold leakage [12.12],[12.13].

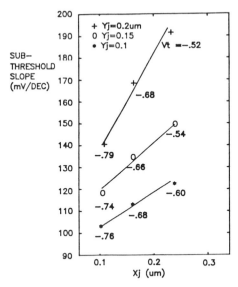

Fig. 12.9. Simulated subthreshold slope vs. X_j and Y_j for p-channel MOSFET. $L_{eff} = 0.5$ μm, $T_{ox} = 25$ nm, $V_{DS} = -3$ V. Vt is the threshold voltage.

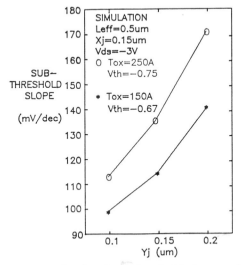

Fig. 12.10. Simulated subthreshold slope vs. Y_j and T_{ox}.

The DIBL effect can also be characterized by the sensitivity of the threshold voltage to the channel length. The reduction of the threshold voltage as the channel length is reduced is due to the potential barrier lowering by the drain bias. The experimental data for p-channel MOSFET with gate oxide thickness of 15 nm are shown in Fig. 12.11. This reduction in threshold voltage may cause circuit performance problems, especially in analog design such as sense amplifiers in memory circuits. It may also cause excessive off current in the transistors, yielding a large standby power. In general, one would like to reduce the sensitivity of the threshold voltage to the effective channel length, thus minimizing the threshold voltage variations throughout the circuit. Fig. 12.12 shows the simulation of this sensitivity for the p-channel transistor with two values of Y_j. It can be seen that the sensitivity increases with increasing Y_j, which implies that Y_j should be minimized as much as possible. The sensitivity will also be reduced if a thinner oxide is used. This is because the channel potential will be more controlled by the gate, thus reducing the drain bias effect. The threshold sensitivity for $T_{ox} = 25$ and 15 nm is shown in Fig. 12.13. The results indicate that thin oxides are essential for very short channel devices.

The leakage current (I_L), defined as the drain to source current at zero gate bias, is simulated for different values of X_j and Y_j. The drain to source bias is -3 V. The results are shown in Fig. 12.14. The interpretation of the graph is the same as for Fig. 12.9. The leakage current is calculated at $X_j = 0.1\ \mu\text{m}$ for $Y_j = 0.1, 0.15$, and $0.2\ \mu\text{m}$, with approximately the same threshold voltage. As the value of X_j is increased, the leakage current increases exponentially. Also, for a fixed X_j (for example, $0.15\ \mu\text{m}$), and for approximately the same V_T, the leakage current increases exponentially with Y_j. The simulated data show that in order for the leakage current to be below 1 nA for a channel length of $0.5\ \mu\text{m}$ and width of $50\ \mu\text{m}$ (this criterion is arbitrary, and is really dependent on the application), X_j and Y_j must be below 0.2 and $0.15\ \mu\text{m}$ respectively. If Y_j can be maintained at $0.1\ \mu\text{m}$, then the criterion for X_j is relaxed to $0.25\ \mu\text{m}$ [12.10]. This simulation is very useful in guiding the development of the submicron CMOS process. It provides the limiting values of the critical parameters such as T_{ox}, X_j and Y_j. These parameters have significant effect on the processing procedures. For example, if a very shallow

Submicron CMOS Technology

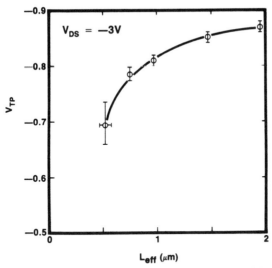

Fig. 12.11. Experimental data of p-channel threshold voltage vs. L_{eff}, for gate oxide thickness of 15 nm.

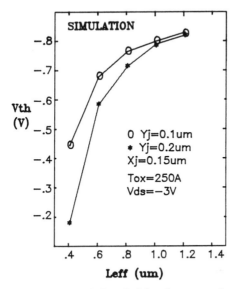

Fig. 12.12. Simulation of threshold voltage vs. L_{eff} for p-channel MOSFET.

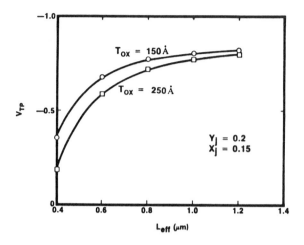

Fig. 12.13. Simulated p-channel threshold voltage vs. L_{eff} for two different gate oxide thicknesses.

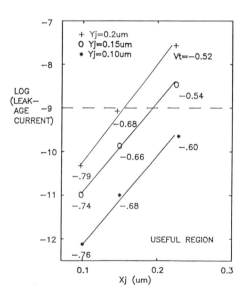

Fig. 12.14. Simulated subthreshold leakage current vs. Y_j and X_j. L_{eff} = 0.5 μm, W = 50 μm, T_{ox} = 25 nm, V_{GS} = 0 V, V_{DS} = -3 V.

source/drain junction is required, this will put a serious constraint on the temperature cycles of the subsequent backend process, such as interlevel dielectric reflow or planarization. It is always necessary to have estimates on the limiting values of these device parameters during process development.

Arsenic Implant Technique to Reduce Y_j

The conventional approach to the counter-doping process for adjusting the p-channel threshold voltage is the implantation of B_{11} at 25 KeV, which produces a junction depth of about 0.18 μm. The simulated profile is shown in Fig. 12.4 in section 12.2. As has been shown, this counter-doping causes severe short channel effects such as subthreshold leakage. A new process to produce a shallow counter-doping junction depth has been developed with the help of simulations. This technique uses an arsenic implant immediately following a BF_2 implant. BF_2 is used because it reduces the channeling effect that is observed in the case of B_{11}. This produces a shallower junction compared with B_{11}. Furthermore, arsenic diffuses relatively slowly during process temperature cycles. It forms a steep profile at the counter-doping junction, and maintains a shallow junction depth. The simulation is shown in Fig. 12.15, and gives a junction depth of 0.09 μm.

One major concern is whether this arsenic implant will introduce a much higher n-type impurity concentration near the surface, which will increase the capacitance between the p^+ diffusion islands and the substrate. As can be seen in the simulation, the n-type impurity concentration is only slightly higher than the conventional technique. Also, the position of the depletion edge under the p^+ diffusion will be beyond the peak of the arsenic profile which only extends to about 0.2 μm. It is clear that this technique will not significantly increase the diffusion parasitic capacitance. One disadvantage of this technique is the process variation sensitivity. The boron channel implant is compensated by an arsenic implant, which requires good process control. Another disadvantage is that this technique requires an extra masking step to prevent the arsenic implant from going into the n-channel transistors. The n-well mask can be used and the alignment is non-critical.

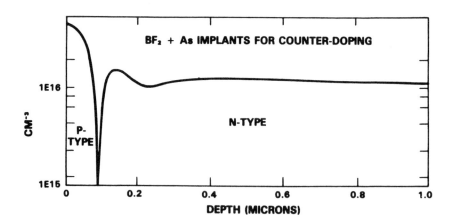

Fig. 12.15. Simulated channel profile for p-channel with BF_2 and As counter-doping.

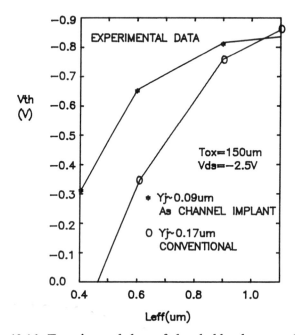

Fig. 12.16. Experimental data of threshold voltage vs. L_{eff} for p-channel with different Y_j.

Submicron p-channel transistors with reduced short channel effects were fabricated using this technique. Fig. 12.16 shows the experimental data of the threshold voltage vs. effective channel length, for a drain to source bias of -2.5 V. The devices with larger values of Y_j showed more threshold roll-off as the channel length is reduced.

N⁻ Pockets for Reducing Short Channel Effects

Another technique for reducing the short channel effects of p-channel transistors is the use of n⁻ pocket [12.14]. This idea is similar to p⁻ pocket for n-channel transistors. Fig. 12.17 shows the transistor structure and potential contours generated by PISCES. (See also Fig. 13.1 in Ch. 13). The n⁻ pockets are produced by implanting phosphorus such that the peak of the impurity profile is above the bottom of the junction. Note that it is very desirable that the n⁻ pocket not to be much deeper than that of the source/drain junction to avoid a significant increase of junction capacitance. The n⁻ impurity concentration is typically 1E17 cm⁻³. As the channel length of the transistor is reduced, the n⁻ pockets of the source and drain are brought closer together, thus effectively raising the n-well impurity concentration. This increasing effective n-well impurity concentration tends to compensate the increasing short channel effects as the channel length is reduced. The idea is to maintain a stable threshold voltage down to L_{eff} = 0.5 µm and also to minimize the subthreshold leakage current. Fig. 12.18 shows the simulation of the subthreshold leakage current for p-channel transistor with and without n⁻ pocket, for L_{eff} = 0.5 µm. The n⁻ pocket design has a much lower subthreshold leakage current. The disadvantage of the n⁻ pocket design is possibly an additional masking step, as well as sensitivity to spacer thickness and n⁻ pocket implant variations. The trade-offs between p-channel transistor with and without the n⁻ pocket is described in detail in Chapter 13, where a novel matrix approach is used to systematically study the sensitivity of the device characteristics to device design parameters.

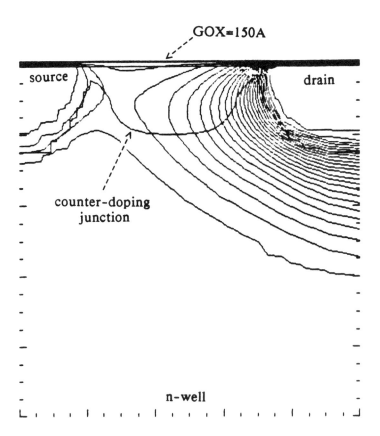

Fig. 12.17. The potential contours of a p-channel MOSFET with n⁻ pockets, generated by PISCES. The gate oxide thickness is 15 nm and effective channel length is 0.5 μm. The drain bias is -5 V and gate bias is 0 V. The interval between contour lines is 0.25 V. Note that the n⁻ pockets reduce the counter-doping junction depth at the source and drain region.

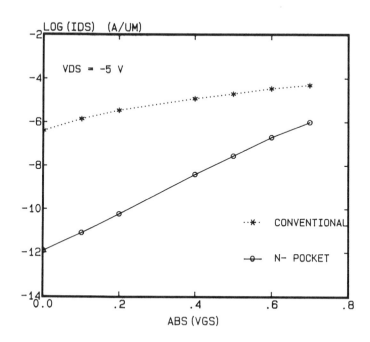

Fig. 12.18. Simulation of the subthreshold characteristics of p-channel MOSFET's with and without n⁻ pockets, using the PISCES program.

12.3 N-Channel Transistor Simulations

Threshold and DIBL are less difficult to control in n-channel MOSFET's than in the p-channel counterpart. The n-channel transistor with n^+ polysilicon gate requires a channel implant of the same type as the substrate, which reduces the short channel effects. Two-dimensional simulations are used to calculate the optimal channel implant profiles. There are other device issues that are more specific to n-channel transistors, which are also discussed in the following sections.

Concerns for Submicron N-Channel MOSFET's

There are two major device issues specific to n-channel transistors, in addition to the threshold voltage and DIBL control which are of general concern for submicron devices. They are the avalanche breakdown between the source and drain [12.6] and device performance degradation due to hot electrons [12.7]. The n-channel device is more susceptible to avalanche breakdown than the p-channel, because the impact ionization coefficient for electrons is much higher than that of the holes [12.15]. Also, for conventional source/drain structures, the n-channel transistor source and drain have more abrupt junctions, which means a higher electric field than the p^+ junctions in the p-channel transistor. The higher ionization coefficient and the higher electric field cause a higher level of substrate current. This current produces a voltage drop across the substrate, which reduces the potential barrier between the source and substrate. Thus the channel current increases and produces more substrate current. This produces a positive feedback action which eventually causes the avalanche breakdown of the device.

Degradation due to hot carrier effects is more serious for the n-channel transistor compared with the p-channel transistor. This is because the electrons in the n-channel device have much higher ionization rate. Thus the understanding of the electric field distribution at the drain region as a function of the drain structure is very important for the n-channel MOSFET. New device structures such as the Lightly-Doped Drain (LDD) structure have been developed to reduce the electric field. The electric field can be simulated by the CADDET or PISCES programs. This is presented in detail in Chapter 9.

Threshold and DIBL Control

The threshold of the n-channel transistor as a function of the channel profile can be simulated by SUPREM for the long channel, and by the combination of SUPREM and GEMINI or PISCES for the short channel transistors. The channel profile will depend on the application of the devices, as well as the structural parameters such as gate oxide thickness. For example, the amount of doping will be less if negative substrate bias is used in the circuit. In the

Submicron CMOS Technology

discussion below, the substrate bias is assumed to be zero, as is the case in most CMOS circuits.

From the discussion in the previous section, it can be seen that for p-channel transistors with channel length of 0.5 μm, a thin gate oxide of 15 nm must be used to reduce the residual leakage current. Hence the channel implant for the n-channel transistor will be based on this gate oxide thickness. Fig. 12.19 shows the channel profile which will produce a threshold voltage of 0.5 V for long channel transistors. But the threshold voltage will be reduced when the channel length is reduced to 0.5 μm, and the drain bias increased. This is studied by coupling the channel profile with the GEMINI program, and specifying the desired drain bias. Fig. 12.20 shows the results of such a simulation. The DIBL effect is evident from the shape of the potential profile which extends from the drain towards the source. The high impurity surface concentration has prevented the punchthrough problem near the surface. This reduces the leakage current under a drain bias of about 3 V, where the potential barrier minimum point is found to be at the surface. At higher drain bias, the punchthrough current path will be through the more lightly doped bulk. The reason for this has been explained in Chapter 8.

The threshold voltage as a function of the effective channel length is simulated for the case of 15 nm gate oxide and a drain bias of 3 V, using the GEMINI program. The result is shown in Fig. 12.21. The threshold roll-off is quite significant at L_{eff} of 0.5 μm, but is still tolerable for most applications. The roll-off can be reduced by using a higher channel implant dose, at the expense of lower current drive, higher junction capacitance and body effect (see Table 7.1 in Chapter 7).

Source/Drain Structures

The source/drain structure of submicron devices is determined by several considerations. The first is the high electric field near the drain. This causes hot carrier degradation problems as well as avalanche breakdown, as mentioned previously. Two types of drain structures are used for reducing the electric field. They are the LDD [12.16] and the double diffused junction [12.17] structures. The double diffused drain structure uses a

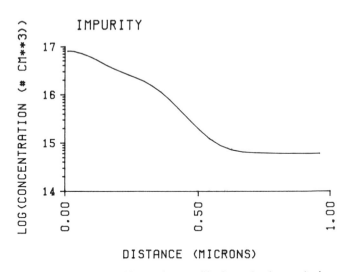

Fig. 12.19. N-channel impurity profile for submicron devices.

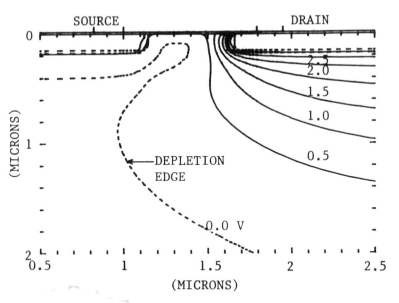

Fig. 12.20. 2-D potential profile of submicron n-channel MOSFET.

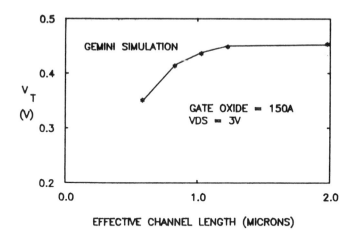

Fig. 12.21. Simulated threshold voltage vs. L_{eff} for n-channel MOSFET.

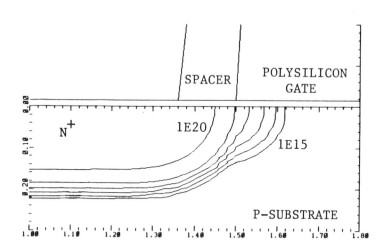

Fig. 12.22. Simulation of the LDD structure by SUPRA.

combination of arsenic and phosphorus implant to form a more graded junction, thus reducing the electric field at the junction. The LDD structure is formed by using a low dose tip implant before the formation of the side-wall spacer, followed by the formation of the spacer and the n^+ diffusion region. Fig. 12.22 shows the simulation of such a structure using the SUPRA program. The tip implant reduces the electric field because of the light doping. The detailed analysis of the LDD structure and its effect on the drain electric field is presented in Chapter 9.

The second consideration in the source/drain structure is process control. For submicron gate length, side-wall spacers are commonly used. This means that either a tip implant is used, or the source/drain region has to be diffused far enough into the channel region to ensure overlap between the gate and the n^+ regions. In this respect, the tip implant is more desirable since the overlap is guaranteed, independent of side-wall spacer thickness variations. Also, the tip junction depth is only about 0.1 μm, which reduces the punch-through problem at short channel lengths, for the same amount of gate to source/drain overlap.

The third consideration is the process complexity. In NMOS technology, the tip implant does not require additional masking. But for CMOS, the tip implant requires an additional mask since the p-channel device has to be masked during the tip implant. The source/drain structure of the submicron n-channel transistor has to be determined by the application of the device. In any case, the conventional arsenic structure will not be adequate due to reliability problems. Either double diffused junction or LDD must be used.

12.4 Summary

The design of the n and p-channel transistors for the submicron n^+ polysilicon gate CMOS process has been presented. The major concern for the p-channel transistor is subthreshold leakage due to the counter-doping. Using simulations, the device can be designed to minimize the short channel effects. For the n-channel transistor, the leakage problem due to short channel effects is not as severe. But the n-channel transistor has more reliability concerns such as the hot electron effects and avalanche breakdown between the source and drain. This requires more complicated source/drain structures such as graded junction and LDD structures. The study of the source/drain structure becomes more important and Chapter 9 deals with this subject in detail using simulations. Computer-aided design has been shown to be essential in advanced device development where device structures are becoming more and more complicated due to many effects related to short channel devices.

References

[12.1] F. Lee, N. Godinho, and C. P. Chiu, "Cool-Running 16K RAM Rivals N-Channel MOS Performance," *Electronics*, Oct 6, 1981, pp. 120-123.

[12.1] Y. El-Mansy, "MOS Device and Technology Constraints in VLSI," *IEEE Trans. on Electron Devices*, **ED-29**, Apr 1982, pp. 567-573.

[12.3] L. C. Thomas and J. J. Molinelli, "VLSI Logic," *Tech Digest of ISSCC 1981*, pp. 230-231.

[12.4] R. R. Troutman, "Recent Developments in CMOS Latchup," *Tech. Digest of IEDM 1984*, pp. 296-299.

[12.5] D. B. Estreich, "The Physics and Modeling of Latch-up in CMOS Integrated Circuits," Stanford Electronics Labs Report #G-201-9, 1980.

[12.6] S. E. Laux and F. H. Gaensslen, "A Study of Avalanche Breakdown in Scaled N-Channel MOSFET's," *Tech. Digest of IEDM 1984*, pp. 84-86.

[12.7] C. Hu, "Hot Electron Effects in MOSFET's," *Tech. Digest of IEDM 1983*, pp. 176-181.

[12.8] K. M. Cham, D. W. Wenocur, J. Lin, C. K. Lau, H.-S. Fu, "Submicrometer Thin Gate Oxide P-Channel Transistors with P+ Polysilicon Gates for VLSI Applications," *IEEE Electron Device Lett.*, **EDL-7**, Jan. 1986, pp. 49-52.

[12.9] S. J. Hillenius, R. Liu, G. E. Georgiou, R. L. Field, D. S. Williams, A. Kornblit, D. M. Boulin, R. L. Johnston, W. T. Lynch, "A Symmetric Submicron CMOS Technology," *Tech. Digest of IEDM 1986*, pp. 252-255.

[12.10] K. M. Cham and S. Y. Chiang, "Device Design for the Submicrometer p-Channel FET with n+ Polysilicon Gate," *IEEE Trans. on Electron Devices*, **ED-31**, July 1984, pp. 964-968.

[12.11] D. V. Morgan, Ed., *Channeling: Theory, Observation and Applications*, New York:Wiley, 1973.

[12.12] T. Kobayashi, S. Horiguchi and K. Kiuchi, "Deep-Submicron MOSFET Characteristics with 5nm Gate Oxide," *Tech. Digest of IEDM 1984*, pp. 414-417.

[12.13] J. R. Brews, W. Fichtner, E. H. Nicollian, and S. M. Sze, "Generalized Guide for MOSFET Miniaturization," *IEEE Electron Devices Lett.,*, **EDL-1**, Jan 1980, pp. 2-4.

[12.14] S. Odanaka, M. Fukumoto, G. Fuse, M. Sasago, T. Yabu, and T. Ohzone, "A New Half-Micrometer P-Channel MOSFET with Efficient Punchthrough Stops," *IEEE Trans. on Electron Devices*, **ED-33**, Mar. 1986, pp. 317-321.

[12.15] E. Takeda, Y. Nakagome, H. Kume, N. Susuki, and S. Asai, "Comparison of Characteristics of n-Channel and p-Channel MOSFET's for VLSI's," *IEEE Trans. on Electron Devices*, **ED-30**, June 1983, pp. 675-680.

[12.16] H. Katto, K. Okuyama, S. Megure, R. Nagai and S. Ikeda, "Hot Carrier Degradation Modes and Optimization of LDD MOSFET's," *Tech. Digest of IEDM 1984*, pp. 774-777.

[12.17] K. Balasubramanyam, M. J. Hargrove, H. I. Hanafi, M. S. Lin, D. Hoyniak, J. LaRue and D. R. Thomas, "Characterization of As-P Double Diffused Drain Structure," *Tech. Digest of IEDM 1984*, pp. 782-785.

Chapter 13
A Systematic Study of Transistor Design Trade-offs

13.1 Introduction

The ability to manufacture circuits at, or near, fundamental density and speed limits is affected by the sensitivity of circuit parameters to variations in the manufacturing process, and by the ability to achieve tight control of the manufacturing process. Modifications of device or circuit design can alter the various dependencies, so that parts of the process that are intrinsically more controllable determine critical circuit properties. Critical dimensions determined by a series of steps (mask-making, exposure, develop, etch) are typically held to +/- 20%. Various ion implant doses and depths can be controlled to +/- 2% to 5%. For example, if there is an absolute minimum for channel length, then the circuit design must be set at $L_{min} + \Delta L$, where the yield of devices with $L > L_{min}$ is satisfactory and ΔL is a measure of the process variation with gate line width. L_{min} must also allow for some overlap of the gate to source and drain. If the minimum channel length is set by a reliability factor, then the design target is determined by allowable early field failures rather than the allowable yield. This illustrates the difficulty in understanding the problems associated with scaling critical dimensions to less than 0.5 μm.

This chapter will not give a framework for solution of all the issues associated with submicron scaling, but present a method to systematically compare the sensitivity of two device designs on various process variables. The devices under study are the p-channel transistor with and without an n⁻ pocket. The results are summarized in a matrix format so that the trade-offs of the two structures can be compared directly. This serves as a good example of how CAD can be used to systematically examine device structures for submicron applications.

13.2 P-Channel MOSFET with N⁻ Pockets

The issues in the design of p-channel transistors with submicron channel length and n^+ polysilicon gate have been described in Chapter 12. The major issue is the subthreshold leakage problem, which is caused by the counter-doping needed to adjust the threshold voltage [13.1]. One proposal for improving this situation involves the use of "n⁻ pockets" [13.2] as shown schematically in Fig. 13.1. These pockets are formed by a relatively deep phosphorus implant before the sidewall spacer oxide is deposited. This extra doping effectively increases the n-well impurity concentration and also reduces the counter-doping depth near the source and drain regions. From the discussion in Chapter 12, the subthreshold leakage is expected to be reduced. Also, as the channel length is reduced, the n⁻ pockets are brought closer together, making the effective n-well concentration higher with reducing channel length. This tends to counteract the threshold voltage fall-off problem.

The SUPREM-PISCES programs have been used to simulate the devices. Fig. 13.2(a) and 13.2(b) show a comparison of the potential distribution for a conventional device and one with the n⁻ pockets. The reduction in the counter-doping junction depth and the depletion depth (the lowest potential contour in the figures) at the source n⁻ pocket region indicates reduced short channel effects.

Design Trade-offs

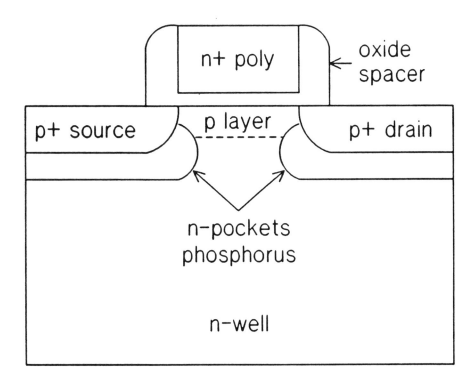

Fig. 13.1. The p-channel MOSFET with n⁻ pockets. The pockets are formed by a phosphorus implant before spacer oxide deposition. The energy of the implant is chosen to position the peak of the phosphorus impurity profile near the p⁺ junction.

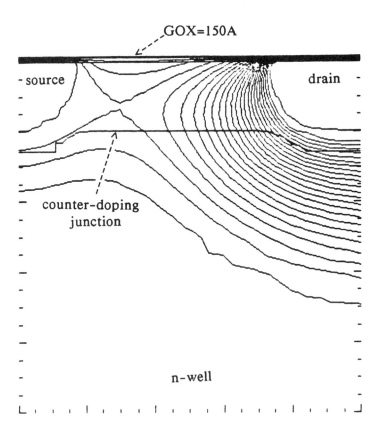

Fig. 13.2(a). Potential contours for a p-channel MOSFET with gate oxide thickness of 15 nm and effective channel length of 0.5 μm. The drain bias is -5 V and gate bias is 0 V. The interval between contour lines is 0.25 V.

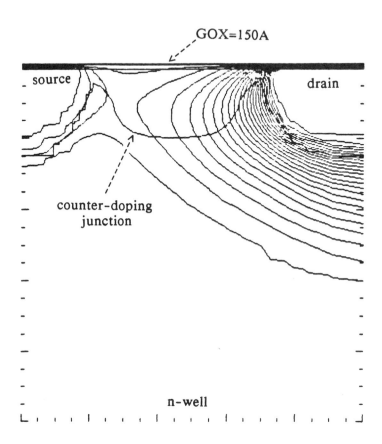

Fig. 13.2(b). Potential contours for the n⁻ pocket p-channel MOSFET. The gate oxide thickness is 15 nm and effective channel length is 0.5 μm. The drain bias is -5 V and gate bias is 0 V. The interval between contour lines is 0.25 V.

13.3 The Sensitivity Matrix

The sensitivity of the device parameters to process variations is of major importance to determine the manufacturability. We will consider the representation of the sensitivity in a general manner and apply this representation to the p-channel device.

A single device parameter, such as leakage current, is a function of various material and structural features such as junction depth, implant dose, etc. These features are, in turn, determined by processing conditions. The process will typically have a nominal value and allowable spread. The spread is typically determined by the ease of control of the equipment and environment as well as the ease of measurement of the feature or features that are most dependent on the process in question.

If all of the process controls are nominal (zero order) then we expect that the resulting features have their nominal values. The device or circuit parameters will also have their nominal values. It should be pointed out that presently available process models are not sufficiently accurate to give an exact relation between process conditions and device features. The approximate relation is generally adequate.

A general device simulator such as PISCES gives a functional connection between structural features and device parameters. We must be concerned with the statistical spread of device parameters from a process. If device features or process controls are chosen to be statistically independent, the analysis is relatively simple assuming that any given feature has a Gaussian probability distribution. That is to say:

$$P(X - \overline{X}) = \frac{1}{\sqrt{2\pi}\,\sigma} \exp\left[-(X - \overline{X})^2 / 2\sigma^2\right] \qquad (13.1)$$

where X is the particular feature (junction depth, implant dose, etc.), $\overline{X}$ is the average or target value, σ is the statistical spread, and $P(X - \overline{X})\Delta X$ is the fraction of events that occur between $X = X - \overline{X}$ and $X = X - \overline{X} + \Delta X$. The fraction of events occurring in the interval of $X = \overline{X} \pm \sigma, 2\sigma, 3\sigma$ is 0.68, 0.95, 0.997 respectively.

If we consider electrical parameter variation around a nominal value, and if the variation is the result of Gaussian deviations of a process around its nominal value, then for small variations the electrical parameter will be Gaussian. If the process deviations are statistically independent, the parametric spread is related in a simple way to the process spread. Thus

$$P(V_i - \overline{V_i}) = \frac{1}{\sqrt{2\pi}\,\sigma_i} \exp[-(V_i - \overline{V_i})^2 / 2\sigma_i^2] \tag{13.2}$$

where V_i is the ith parameter (voltage, current, etc.), $\overline{V_i}$ is the nominal value, σ_i is the ith statistical spread.

$$\sigma_i^2 = \sum (A_{ij}\sigma_j)^2 \tag{13.3}$$

where σ_j is the statistical spread of the jth process feature, and A_{ij} is the sensitivity of the parameter $V_i - \overline{V_i}$ to the jth process feature.

To compare design differences of p-channel transistors with and without n⁻ pockets, ten process parameters were considered and five device characteristics were extracted. The target or nominal values of these parameters are listed in Table 13.1(a) and 13.1(b). The ten process features and 5 device parameters results in a 5 × 10 matrix of process sensitivity.

The exact representation of device parameter depends somewhat on the particular parameter. Log(I_L) is more convenient to use for leakage current than the current itself because the spread of current may be several orders of magnitude. The linear sensitivity is accurate for only a limited range of process variables. We chose to represent the process variables as fractional deviations from the target value. Fig. 13.3(a) and 13.3(b) show the calculated matrix A_{ij} and its connection between process and device for the n⁻ pocket and conventional source/drain devices respectively. Note that within each matrix, there is no meaning in comparing the magnitude of the matrix elements, since they are dependent on the definitions of the process and device parameters. It is only meaningful to compare the same matrix element between the matrices in Fig. 13.3(a) and 13.3(b).

To extract further information from this matrix, it is necessary to analyze a given process environment so that each of the statistical variations can be estimated. For our purpose it is sufficient to know that the percentage spread on all of the processes except gate length is less than 10%. For short channel length, the control of gate length may easily be no better than 20%. If these

N⁻ Pocket Devices

Process Parameters:

N-well doping (N_W) = 1E16 cm^{-3}

P$^+$ junction depth (X_j) = 0.25 µm

Counter-doping junction depth (Y_j) = 0.15 µm

Counter-doping dose (D_C) = 9.3E11 cm^{-2}

Gate oxide thickness (T_{ox}) = 15 nm

Spacer thickness (S_p) = 0.16 µm

N⁻ pocket dose (D_n) = 3.5E12 cm^{-2}

N⁻ pocket peak depth (R_n) = 0.2 µm

N⁻ pocket characteristic length (C_n) = 0.1 µm

Poly gate length (L_p) = 0.55 µm

Device Parameters:

Subthreshold leakage current $(\text{Log}(I_L))$ = -11; (A/µm, V_D = -3 V)

Low drain bias threshold voltage $(V_{T(LOW)})$ = -0.71 V; $(V_D$ = -0.05 V)

High drain bias threshold voltage $(V_{T(HI)})$ = -0.5 V; $(V_D$ = -3.3 V)

Drain Capacitance (C_D) = 0.52 fF/µm^2

Long channel threshold voltage (V_{TLC}) = -0.62 V

Table 13.1(a). Process and device parameters for n⁻ pocket p-channel device.

Design Trade-offs

Conventional Source/Drain Devices

Process Parameters:

N-well doping (N_W) = 4E16 cm^{-3}

P$^+$ junction depth (X_j) = 0.20 μm

Counter-doping junction depth (Y_j) = 0.13 μm

Counter-doping dose (D_C) = 1.32E12 cm^{-2}

Gate oxide thickness (T_{ox}) = 15 nm

Spacer thickness (S_p) = 0.13 μm

Poly gate length (L_p) = 0.55 μm

Device Parameters:

Subthreshold leakage current $(Log(I_L))$ = -11; (A/μm, V_D = -3 V)

Low drain bias threshold voltage $(V_{T(LOW)})$ = -0.72 V; (V_D = -0.05 V)

High drain bias threshold voltage $(V_{T(HI)})$ = -0.53 V; (V_D = -3.3 V)

Drain Capacitance (C_D) = 0.40 fF/μm^2

Long channel threshold voltage (V_{TLC}) = -0.80 V

Table 13.1(b). Process and device parameters for conventional p-channel device.

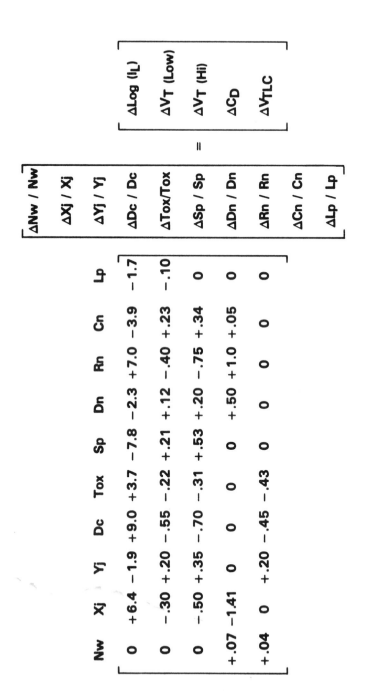

Fig. 13.3(a). Sensitivity Matrix for n⁻ pocket p-channel device.

$$
\begin{bmatrix}
& Nw & Xj & Yj & Dc & Tox & Sp & Dn & Rn & Cn & Lp \\
& -3.7 & +5.8 & -6.5 & +17 & +3.9 & -4.4 & 0 & 0 & 0 & -9.2 \\
& +.20 & -.12 & +.65 & -.90 & -.30 & +.05 & 0 & 0 & 0 & +.11 \\
& +.32 & -.33 & +.70 & -1.5 & -.35 & +.25 & 0 & 0 & 0 & +.54 \\
& +.17 & 0 & 0 & 0 & 0 & 0 & 0 & 0 & 0 & 0 \\
& +.20 & 0 & +.73 & -.71 & -.21 & 0 & 0 & 0 & 0 & 0 \\
\end{bmatrix}
\begin{bmatrix} \Delta Nw/Nw \\ \Delta Xj/Xj \\ \Delta Yj/Yj \\ \Delta Dc/Dc \\ \Delta Tox/Tox \\ \Delta Sp/Sp \\ \Delta Dn/Dn \\ \Delta Rn/Rn \\ \Delta Cn/Cn \\ \Delta Lp/Lp \end{bmatrix}
=
\begin{bmatrix} \Delta \mathrm{Log}(I_L) \\ \Delta V_T \text{ (Low)} \\ \Delta V_T \text{ (Hi)} \\ \Delta C_D \\ \Delta V_{TLC} \end{bmatrix}
$$

Fig. 13.3(b). Sensitivity Matrix for conventional p-channel device.

estimated values for process control are used, then the statistical spread of device parameters can be estimated from the sensitivity matrix. The following shows the estimated 99.7% spread ("3-sigma" variations) of the device characteristics with/without n⁻ pocket (units same as in Table 13.1(a),(b)):

$$\sigma(Log(I_L)) = 1.7 / 2.7$$

$$\sigma(V_{TLOW}) = 0.1 / 0.1$$

$$\sigma(V_{THIGH}) = 0.14 / 0.2 \tag{13.4}$$

$$\sigma(\Delta C_D) = 0.2 / 0.017$$

$$\sigma(\Delta V_{TLC}) = 0.2 / 0.1$$

The indicated device parameter spreads show some differences between the n⁻ pocket and conventional source/drain structure. The n⁻ pocket device parameters has less variations due to short channel effects, but has more variations on the junction capacitance and long channel threshold voltage. The sensitivity matrix shows specifically the process variable that is the source of parameter variation. The comparison may be summarized as:

Advantages of n⁻ pocket:

1. Nearly eliminates the dependence of leakage and threshold voltage on gate length. This allows the long channel threshold voltage to be reduced from -0.8 V to -0.6 V.

2. Greatly reduces the dependence of subthreshold leakage and threshold on counter-doping dose and depth.

3 May be scalable to channel length less than 0.5 μm.

Disadvantages:

1. Subthreshold leakage and threshold voltage become more sensitive to spacer width and p⁺ junction depth. These two parameters determine the size of the n⁻ pockets.

2. Process complexity is increased due to extra implant and perhaps one extra masking step. Subthreshold leakage and threshold voltage now depend on n^- pocket implant parameters.

3. Drain capacitance becomes sensitive to the n^- pocket implant parameters and p^+ junction depth.

13.4 Conclusions

By using the SUPREM-PISCES programs, the sensitivity of the device electrical characteristics to process parameters variations can be systematically studied. In this particular case study, the sensitivity matrix shows that while the n^- pocket design reduces the short channel problems, the structure introduces sensitivities to new processing variables. The desirability of using one or the other design depends on the controllability of various processes.

References

[13.1] K. M. Cham and S. Y. Chiang, "Device Design for the Submicrometer p-Channel FET with n^+ Polysilicon Gate," *IEEE Trans. on Electron Devices*, **ED-31**, July 1984, pp. 964-968.

[13.2] S. Odanaka, M. Fukumoto, G. Fuse, M. Sasago, T. Yabu, and T. Ohzone, "A New Half-Micrometer p-Channel MOSFET with Efficient Punchthrough Stops," *IEEE Trans. Electron Devices*, **ED-33**, March 1986, pp. 317-321.

Chapter 14
MOSFET Scaling by CADDET

14.1 Introduction

The density and performance of integrated circuits have increased by many orders of magnitude through the process of device scaling. As pointed out in the overview chapter, the long channel relations are not strictly valid for horizontal dimensions that are comparable to the vertical dimensions. In this example, the operating voltage is kept constant. The horizontal dimension, L_{eff}, and the vertical dimension, T_{ox}, will be separately scaled to approximately two-thirds of the established process values. The two scaling factors are not identical, but are in the typical scaling range. The result of this reduction is then established. Comparison with the long-channel scaling assumptions is possible, and the importance of secondary physical effects can be seen. Device width will not be scaled, the current drive capability will be expressed for a fixed width. Breakdown and punchthrough voltages must be sufficiently greater than the supply voltage so that reliability is not a problem. For this example, power density is not a limitation.

Two different modes of devices are considered, an enhancement mode device and a depletion mode device. Section 14.2 discusses how device simulation programs are used to study an enhancement mode device scaling and its

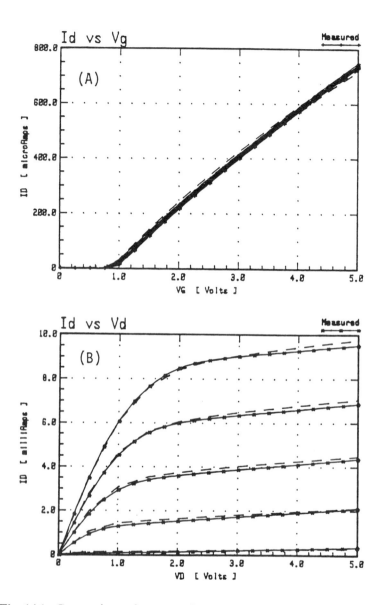

Fig. 14.1. Comparisons between the measurements (solid lines) and CADDET simulations (dashed lines). (A): V_B = -2 to -5 V, (B): V_G = 1 to 5 V, V_B = 0. Device: T_{ox} = 40 nm, L_{eff} = 1.25 μm, W_{eff} = 48.25 μm.

MOSFET Scaling

results are compared to the classical predictions [14.1]. Next, a scaling scheme of a depletion mode device is considered in Section 14.3 and is contrasted to the enhancement mode device scaling. Section 14.4 presents conclusions. The device simulation program CADDET is used throughout for numerical calculations.

14.2 Scaling of an Enhancement Mode MOSFET

CADDET Simulation

A typical n-channel enhancement mode device is chosen as a standard device to develop the discussion. The mask dimensions of the gate are 50 μm wide and 2 μm long. Structural data are collected by a series of measurements. The gate oxide thickness is 40 nm measured by C-V technique. [14.2] The measured effective channel length (L_{eff}) and the source-drain series resistance are 1.25 μm and 750 Ω per μm width respectively [14.3]. The electrical characteristics of the device are shown in Fig. 14.1. The measured data are corrected for the voltage drop across the source-drain series resistances.

The electrical characteristics of short channel length devices deviate from the classical model and do not conform to any known simple analytic expressions [14.4]. Numerical process and device simulation programs become very useful tools to study the characteristics of small geometry devices due to the built-in generality of the device structure and the accuracy in numerical computation.

CADDET is a 2-D device characteristics simulation program developed by HITACHI. The program solves Poisson's equation in the substrate with the current continuity equation for a single carrier only. It assumes that the surface of the substrate is planar and that the gate covers the entire substrate surface. In our application, the electron mobility in the program is replaced by a new expression given by Eq. (3.15). The substrate doping profile is simulated by running SUPREM [14.5], a 1-D process simulation program, which gives the doping profile in the substrate when the process sequences are specified (see Ch. 2). The source/drain profiles are also obtained from SUPREM.

	SUPREM Results		
	Channel Implant	Source/Drain Implant	Depletion Implant
Element	Boron	Arsenic	Arsenic
Dosage (cm^2)	5.3E11	5.0E15	1.8E12
Peak Concentration (cm^3)	2.65E16	6.6E20	1.67E17
Peak Position (R_P) (μm)	0.12	0.00	0.07
Standard Deviation (ΔR_P) (μm)	0.083	0.06	0.043
Junction Depth (x_j) (μm)	-	0.29	0.16

Table 14.1. Conversion of SUPREM results to Gaussian formula.

Since CADDET takes only analytic expressions (such as normal distribution) for doping profiles, the results of SUPREM are converted to a normal distribution as shown in Table 14.1. Based on this data, the current-voltage (I-V) characteristics of the present (unscaled) enhancement mode device are simulated by CADDET and are compared with the measured data in Fig. 14.1. The device parameters that are important for digital circuits are listed in Table 14.2.

W/L Masked (μm/μm)	50/2	50/2	50/1.5	50/1.5
W_{eff} (μm)	48.25	48.25	48.25	48.25
L_{eff} (μm)	1.25	1.25	0.75	0.75
T_{ox} (nm)	40	25	25	25
C_{ox} (F/cm^2)	8.63E-8	1.38E-7	1.38E-7	1.38E-7
Q_{NA} (cm^{-2})	5.3E11	9.5E11	1.17E12	1.17E12
Energy (KeV)	50	50	50	80
V_T (V) at $V_B = 0$ V	0.66	0.59	0.62	0.39
V_T (V) at $V_B = -2$ V	0.90	0.90	0.90	0.80
μ_0 (cm^2/V-sec)	630	580	590	590
I_D (mA) @ $V_{BS} = -2$ V, $V_{GS} = V_{DS} = 5$ V	9.55	13.9	17.4	18.4
Current Increase due to Scaling	1.0	1.46	1.82	1.92

Table 14.2. Device parameters for unscaled and scaled enhancement mode devices.

Scaling of Gate Oxide Thickness

As a first step in scaling, the gate oxide thickness (T_{ox}) is reduced from 40 nm to 25 nm. This gives a scaling factor of 1.6 in T_{ox}. The reduction in T_{ox} increases the gate capacitance (C_{ox}), which in turn increases the transconductance (g_m) and the drain current. According to the classical model, the drain current is proportional to this scaling factor [14.6].

The threshold voltage (V_T) depends on the implantation dosage to the first order, when the depletion boundary extends below the implantation region. CADDET simulations are done to study V_T variation with implantation dosage for a fixed energy of 50 KeV. The boron implantation dose of 9.5E11 cm^{-2} was chosen to keep V_T at 0.9 V. This process is designed to use a substrate bias of -2 V and thus all threshold voltages refer to this bias condition unless otherwise stated.

I-V characteristics of the device with the scaled T_{ox} and the adjusted implantation are shown in Fig. 14.2. The important parameters of these characteristics are listed in Table 14.2. They show that the reduction in T_{ox} causes a reduction in mobility of 8 %. This is due to the fact that the vertical electric field at the surface degrades the electron mobility. With a fixed gate bias, the thinner the insulator, the higher the electric field at the surface and the smaller the surface electron mobility. Fig. 14.2 shows that the saturation current at $V_{GS} = V_{DS} = 5.0$ V and $V_{BS} = -2$ V is 14 mA. For the same bias condition, the unscaled device had a saturation current of 9.5 mA. Thus the current increases by a factor 1.47 in the scaled device. This number is close to the product of the scaling factor of T_{ox} ($= 1.6$) and the reduction of mobility ($= 0.92$).

It should be noted that the increase in drain current at this stage of scaling does not always lead to an increase in circuit speed. The premise is true when the load of an inverter stage is dominated by parasitic capacitances, thus the increase in the gate capacitance does not affect the total load capacitance very much. So, for fixed load, the charging time is inversely proportional to the driving current. However, in case of a circuit whose load is dominated by gate capacitance, the scaling in T_{ox} alone slows down the circuit speed, because the drain current increases by factor of only 1.47 while the load capacitance

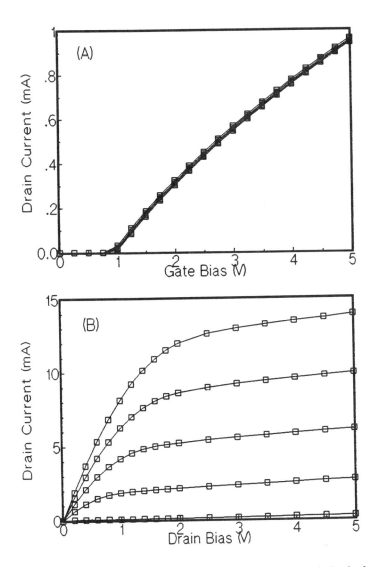

Fig. 14.2. CADDET Simulations of an enhancement mode device with $T_{ox} = 25$ nm, $W_{eff} = 48.25$ μm and $L_{eff} = 1.25$ μm. (A): $V_{BS} = -2$ to -5 V; (B): $V_{GS} = 1$ to 5 V ($V_{BS} = -2$ V).

increases by 1.6 in our case. Thus the load increases by a higher factor than the drain current. The present scaling in T_{ox} should be followed by scaling in channel length, which requires readjustment of implantation dose and energy.

Scaling of Channel Length

With the gate oxide thickness scaled as discussed in the previous section, the mask channel length (L) is reduced from 2.0 μm to 1.5 μm. Assuming that the source/drain regions are formed by the same process technology before and after the scaling, this scaling reduces the effective channel length (L_{eff}) from 1.25 μm to 0.75 μm. This results in a scaling factor of 1.67. Since L_{eff} is less than one micron, the effects of reducing L_{eff} on the threshold voltage and on the channel current are expected.

With the scaled structure (T_{ox} = 25 nm and L_{eff} = 0.75 μm), the shift of V_T by implantation dosage (Q_{NA}) is simulated at 50 KeV. The implantation dosage of 1.17E12cm^{-2} is chosen so that V_T for V_{BS} = -2 V remains at 0.9 V. CADDET simulation is used to determine the device parameters which are listed in Table 14.2. The mobility is not affected very much by the channel length scaling. The simulated drain current at V_{DS} = V_{GS} = 5 V and V_{BS} = -2 V is 17.4 mA. This is increased by a factor of 1.25 over that of the device with scaled oxide thickness but unscaled channel length. This is less than the scaling factor of channel length (1.67). The drain current for V_{GS} = 5 V and V_{DS} = 0.1 V increased by a factor of 1.59 which is close to the channel length scaling factor. Thus the saturation current increase of only 1.25 must be due to carrier velocity saturation in the channel. The resultant gain in saturation current for scaling both oxide thickness and channel length is 1.84, far less than the product of the two scaling factors (2.67) as predicted by the classical model.

With this fixed implantation dosage, the implantation energy is varied to study the subthreshold current and the punchthrough voltage. First, the subthreshold currents (I_D at V_D = 0.1 V, V_{BS} = -2 V) are simulated in Fig. 14.3 with implantation energy ranging from 20 KeV to 50 KeV. The figure shows that the slope of the current is fixed at 74 mV/decade regardless of the energy.

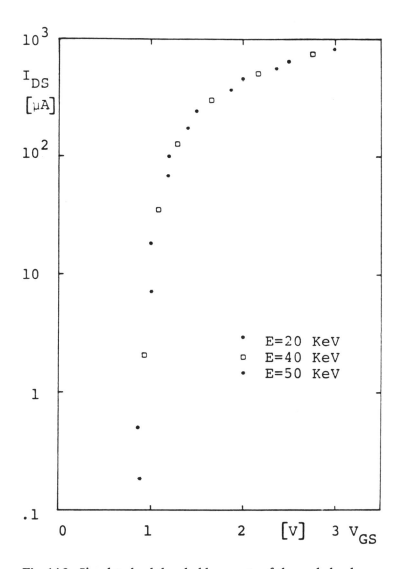

Fig. 14.3. Simulated subthreshold currents of the scaled enhancement mode device. W_{eff} = 48.25 μm, L_{eff} = 0.75 μm, T_{ox} = 25 nm, V_{DS} = 0.1 V, V_{BS} = -2 V, Q_{NA} = 1.17E12 cm^{-2}.

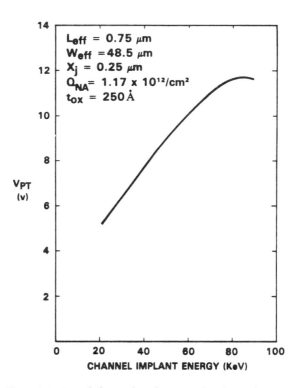

Fig. 14.4. Punchthrough voltage vs. implantation energy.

The implantation energy determines the position of the peak impurity concentration and the punchthrough current path. Fig. 14.4 shows the change of punchthrough voltage (V_{PT}) as a function of implantation energy at fixed dosage. V_{PT} is defined as a drain bias voltage at which the drain current is 1 nA per 1 μm width at V_{GS} = 0 V and V_{BS} = -2 V. The figure shows that V_{PT} is sensitive to energy and has a maximum value at 80 KeV. Above this value, V_{PT} deteriorates due to the occurrence of surface punchthrough. Below this value, bulk punchthrough is the dominant current mechanism. It is known that the punchthrough voltage is very sensitive to the doping profile in the substrate. Our measurements show that the boron implantation follows the Pearson IV distribution rather than the normal distribution. If CADDET were upgraded to take different types of impurity distributions, the accuracy of the punchthrough simulation would improve.

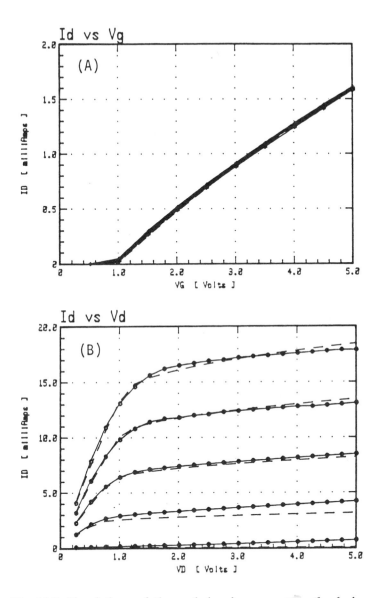

Fig. 14.5. Simulations of the scaled enhancement mode device. T_{ox} = 25 nm and L_{eff} = 0.75 μm. The solid lines are by CADDET and the dashed lines are by SPICE. (A): V_B = -2 to -5 V; (B): V_G = 1 to 5 V.

The I-V characteristics of the scaled structure are shown in Fig. 14.5. The device parameters are extracted from these simulated curves and tabulated in Table 14.2 along with the previous results. The increase of implantation energy to 80 KeV decreases V_T for V_{BS} = -2 V to 0.8 V from its target value of 0.9 V at 50 KeV. The high implantation energy pushes the peak of the impurity profile deeper from the surface and the effective surface concentration decreases, consequently the threshold voltage decreases. The 0.1 V decrease in threshold voltage causes the linear and saturation currents to increase about 6%.

14.3 Scaling of a Depletion Mode MOSFET

CADDET Simulation

A depletion mode device fabricated on the same wafer as the enhancement mode device is chosen as a standard device. The mask dimensions of the gate are 3 μm wide and 50 μm long. The structural data of the device are the same as those of the enhancement mode device. The electrical characteristics of the device are shown in Fig. 14.6. The measured data are corrected for the voltage drop across the source and drain resistances. The difference in g_m between V_{GS} < 0 V and V_{GS} > 0 V indicates the device has a buried channel.

To get the doping profile in the substrate, SUPREM is first run. The profile is converted to a sum of two normal distributions, one for the channel implantation and another for the depletion implantation. Then a CADDET input file is generated to simulate the device. The flat-band voltage (V_{fb}) is adjusted to give the correct magnitude of the drain current at V_{GS} = 0 V. Once this adjustment is made, then the simulation of I-V characteristics shows good agreement with measurement in Fig. 14.6.

The scaling of the depletion mode device depends on the scaling result of the enhancement mode device, since the former is used as a load for the latter. In the following, the scaling scheme of a depletion mode device is discussed following the same steps as an enhancement mode device; the reduction in T_{ox}, followed by the reduction in L_{eff}. In each section, it is pointed out how the

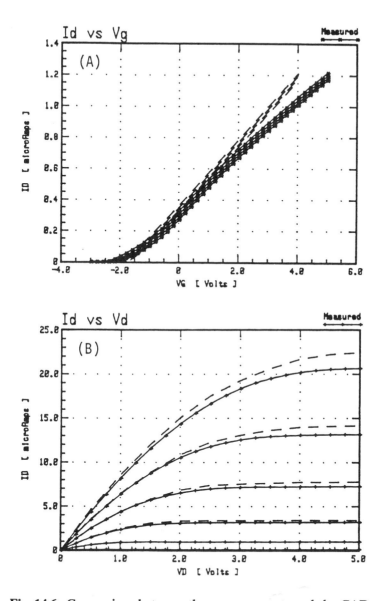

Fig. 14.6. Comparison between the measurements and the CAD-DET simulations of a depletion mode device. T_{ox} = 40 nm, W = 3 μm and L = 50 μm. The solid lines are measurements and the dashed lines simulations. (A): V_{BS} = -2 to -5 V; (B): V_{GS} = -2 to 3 V.

scaling scheme for a depletion mode device is different from the scheme for an enhancement mode device.

Scaling of Gate Oxide Thickness

The current in the buried channel device in the linear region is closely proportional to the amount of the carrier charge in the channel, which is approximately given by

$$qQ_n \approx q(Q_{ND} - Q_{NA}) + C_{ox}(V_{GS} - V_{fb}) \qquad (14.1)$$

$$\text{when} \quad V_{GS} \approx V_{fb}$$

where Q_n is the electron density per unit area. Q_{ND} and Q_{NA} are the dosages of depletion and channel implantations per unit area. The second term in the equation represents the accumulation charge when V_{GS} is greater than V_{fb}. The equation is valid when the depletion boundary lies deeper than the depth of the implantations and $Q_{ND} >> Q_{NA}$. Eq. (14.1) indicates that the charge in the channel is determined by the net implantation dosage and is independent of C_{ox} when $V_{GS} = V_{fb}$. In case of n$^+$ poly gate device, the flat-band voltage (V_{fb}) is determined by

$$V_{fb} = -\frac{kT}{q}\ln\left(\frac{N^+}{N_S}\right) \approx -0.15\,V \qquad (14.2)$$

when the voltage drop in the oxide due to the surface state charge (Q_{ss}) is negligible. N^+ in the equation is the doping concentration in the poly gate and N_S is the surface concentration in the substrate. Hence the magnitude of the V_{fb} of an n$^+$ poly gate depletion mode device is close to zero volt.

In real digital circuits, depletion mode devices are used almost exclusively with zero volt from gate to source. Hence the device operates practically at flat-band condition. Then, the parameters of importance are the current characteristics at zero gate to source voltage, which are governed by the dosages of implantations, and its sensitivity to substrate bias. Eq. (14.1) can be rearranged to

$$qQ_n \approx C_{ox}(V_{GS} - V_T) \quad \text{when} \quad V_{GS} > V_{fb} \qquad (14.3)$$

$$\text{where} \quad V_T = V_{fb} - \frac{q(Q_{ND} - Q_{NA})}{C_{ox}} \tag{14.4}$$

The equation shows that a depletion mode device may be considered like an enhancement mode device when $V_{GS} > V_{fb}$ if we define an extrapolated threshold voltage given by Eq. (14.4). In order to scale the drain current (controlled by Q_{ND} and Q_{NA}) by the same T_{ox} scaling factor as the enhancement mode device, the extrapolated threshold voltage should remain fixed throughout the scaling procedure. In this case, the gate oxide thickness of the depletion mode device is reduced to 25 nm as a result of the enhancement mode device scaling. Also the channel implantation dosage and the energy are set to 1.17E12 cm^{-2} and 80 KeV respectively to optimize the enhancement mode device.

CADDET simulations are performed to study the extrapolated threshold shift by Q_{ND}. The final value of 2.74E12 cm^{-2} at 145 KeV is determined for the reduced gate oxide thickness. With the reduced T_{ox} and readjusted Q_{ND}, CADDET is run to get the electrical characteristics as shown in Fig. 14.7. The current path in a depletion mode device is through a buried channel in which the vertical electric field is virtually zero. Mobility degradation by the vertical field, which was a degradation factor in the scaled enhancement mode device does not show up in the scaled depletion mode device. Thus the drain current increases linearly with the scaling factor.

Scaling of Channel Length

The size of the depletion mode device working as a pull-up transistor in an inverter is determined by the size of the enhancement mode device in the circuit and by the specifications for the logic levels of the digital circuit. Since the logic levels and the enhancement mode device are not specified in our example, the mask channel length is varied from 50 μm to 3.5 μm. The drain currents are simulated by CADDET and the results are shown in Fig. 14.8. If we consider the variation of the saturation current by the channel length, it shows that the current increase of a depletion mode device scaled by the channel length is greater than its scaling factor. For example, the saturation current (I_{DS} at $V_{GS} = 0V$, $V_{DS} = 5$ V) increases from 5.93 μA to 136 μA, a factor of

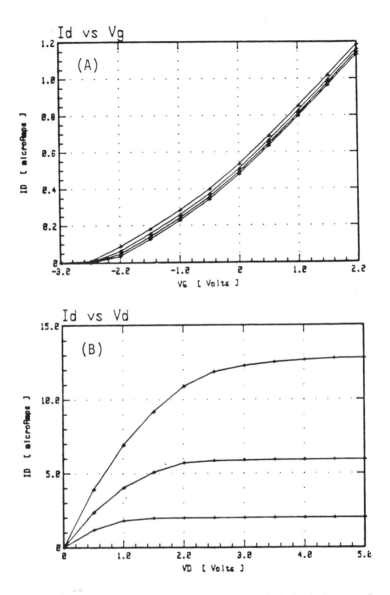

Fig. 14.7. CADDET simulations of the scaled depletion mode device with T_{ox} = 25 nm, L = 50 μm and W = 3 μm. (A): V_{BS} = -2 to -5 V; (B): V_{GS} = -1 to 1 V.

MOSFET Scaling

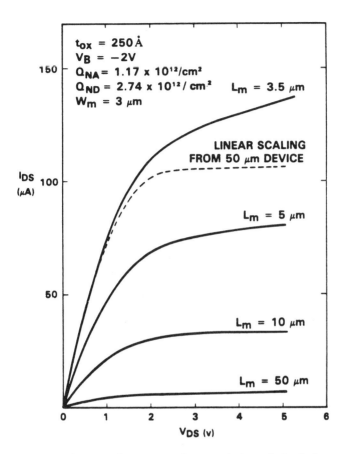

Fig. 14.8. Change of current characteristics of depletion mode devices due to change in channel length, with $V_{GS} = 0$ V.

22.9, when $L_{\it eff}$ is reduced from 49.25 μm to 2.75 μm, a factor of 17.9. The same trend is verified in the experimental measurement. For the case of $L_{\it eff}$ = 2.75 μm, the curve of $L_{\it eff}$ = 49.25 μm is multiplied by the scaling factor (= 49.25/2.75) and plotted in Fig. 14.8. It shows that the current in the linear region scales linearly, while the current in the saturation region has high output conductivity. The excessive slope in the saturation region adds extra transconductance at V_{DS} = 5 V. This extra gain in the current by the $L_{\it eff}$ scaling is opposite to the case of an enhancement mode device. The enhancement mode

W/L Masked (μm/μm)	3/5	3/5	3/3
W_{eff} (μm)	1.5	1.5	1.5
L_{eff} (μm)	4.25	4.25	3.25
T_{ox} (nm)	40	25	25
C_{ox} (F/cm^2)	8.63E-8	1.38E-7	1.38E-7
Q_{NA} (cm^{-2})	5.3E11	1.17E12	1.17E12
Energy (KeV)	50	80	80
Q_{ND} (cm^{-2})	1.8E12	2.74E12	2.74E12
Energy (KeV)	145	145	145
V_T (V) at V_B = -2 V	-1.7	-1.6	-1.6
I_D (μA) @ V_{BS} = -2 V V_{GS} = 0 V_{DS} = 5 V	48.7	80.4	136
Current Increase due to Scaling	1.0	1.65	2.79

Table 14.3. Device parameters for unscaled and scaled depletion mode devices.

short channel device shows mobility degradation in the linear region and considerable velocity saturation in the saturation region. All these effects combined cause the current gain factor of an enhancement mode device to be smaller than the L_{eff} scaling factor. The device parameters of the unscaled and the scaled depletion devices are shown in Table 14.3.

14.4 Conclusions

CADDET is used intensively along with SUPREM to study the scaling schemes for MOSFET devices. It is shown that the MOSFET devices can be scaled to improve device performance. In case of an enhancement mode device, gate oxide thickness and channel length are chosen as prime scaling factors. The reduction in the gate oxide thickness increases the drain current, but the surface mobility degrades and this causes deviation from the linear scaling. The reduction in channel length promotes short channel effects, especially the saturation of the carrier velocity. After the punchthrough voltage is optimized, the final gain in the saturation current is 1.92.

In case of a depletion mode device, the depletion implantation dosage and the channel length become the major scaling factors. Since the current path is through the buried channel, the mobility is insensitive to the vertical field from the gate terminal. However, the excessive output conductance in the saturation region of the reduced channel length should be carefully controlled.

References

[14.1] R. H. Dennard et al, "Design of Ion-Implanted MOSFET's with Very Small Physical Dimensions," *IEEE J. Solid-State Circuits*, **SC-9**, No.5, 1974, pp 256.

[14.2] E. H. Nicollian and J. R. Brews, *MOS Physics and Technology*, New York: John Wiley & Sons, Inc., 1982.

[14.3] R. C. Y. Fang, R. D. Rung and K. M. Cham, "An Improved Automated Test System for VLSI Parametric Testing," *IEEE Trans. Instru. Meas.*, **IM-31**, no. 4, Sept 1982, pp. 198-205.

[14.4] L. A. Akers and J. J. Sanchez, "Threshold Voltage Models of Short, Narrow and Small Geometry MOSFET's: A Review", *Solid-State Electronics*, **25**, No.7, pp 621-641, July 1982.

[14.5] D. A. Antoniadis, S. E. Hansen, and R. W. Dutton, "SUPREM II - A Program for IC Process Modeling and Simulation," TR 5019.2, Stanford Electronics Laboratories, Stanford University, Calif., June 1978.

[14.6] A. S. Grove, *Physics & Technology of Semiconductor Devices*, New York: John Wiley & Sons, Inc., 1967.

[12.7] L. D. Yau, " A Simple Theory to Predict the Threshold Voltage of Short-Channel IGFET's", *Solid-State Electronics*, **17**, pp. 1059-1063, Oct 1974.

Chapter 15
Examples of Parasitic Elements Simulation

15.1 Introduction

In this chapter, the examples of parasitic component simulations are presented. The problems are solved by the appropriate simulation tools discussed in Chapter 4 and the other tools such as SUPREM, GEMINI, and SUPRA. Section 15.2 covers two-dimensional problems and section 15.3 deals with the problems which have to be solved by three-dimensional simulation. The experimental verification of the simulation results are also discussed.

15.2 Two-Dimensional Parasitic Components Extraction

Interconnect Capacitance Calculation by SCAP2

The areal and peripheral coefficients of parasitic capacitances between conductive layers can be investigated using the SCAP2 program. The structures range from the simple case of conducting lines over field oxide to more complicated cases of multi-layer coupling and interline coupling. Examples are:

(1) metal lines over field oxide, polysilicon layers, or diffusion
(2) polysilicon lines over field oxide
(3) polysilicon lines over field oxide covered by metal plane
(4) interline coupling of metal lines over polysilicon plane

Since SCAP2 does not accept semiconductor layers, the silicon layers are assumed to be conductors. This is a reasonable assumption for case (1), where the dielectric thickness is much larger than the depletion layer thickness in the silicon substrate and the bottom plane has high conductivity [15.1],[15.2]. However, when the polysilicon line goes over the field oxide (case (2) and (3)), it forms a depletion layer on the surface of the substrate, the width of which is dependent on the impurity profile under the field oxide, and can be a significant fraction of the field oxide thickness. The GEMINI program is used for this case. This technique will be discussed in detail.

The simplest calculation would be that of a conducting line over a conducting plane (such as diffusion, or silicon substrate) separated by a layer of dielectric such as silicon dioxide, representing case (1). Fig. 15.1 shows such a simulation. The equipotential lines are shown (the electric field lines are perpendicular to the potential contours), which clearly show the importance of the fringing component, i.e., the component of capacitance arising from the two sides of the conducting line. For VLSI, the dielectric structure can be more complicated, and may contain a combination of different dielectric materials. This is due to issues such as planarization for better line width control. SCAP2 can also simulate these cases. The areal and fringing components of the capacitance are extracted by performing the simulation with different widths of conducting line. By plotting the total capacitance versus the line width, the slope of the resulting line and the extrapolation to y-axis will provide the areal and fringing components of the parasitic capacitance, respectively. Fig. 15.2 shows the result for the case where the interlevel dielectric is a combination of nitride and oxide.

As an example of the study of a more complicated structure, SCAP2 is used to simulate case (3) although the approximation that the substrate is metallic has to be made. Fig. 15.3 shows the structure generated by SCAP2. The program calculates the charge induced on the metal and substrate, hence

Examples of Parasitics Simulation

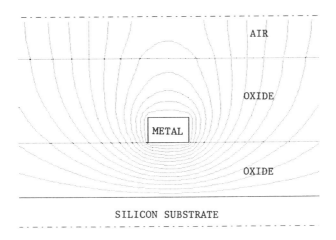

Fig. 15.1. SCAP2 simulation of a metal line over field oxide.

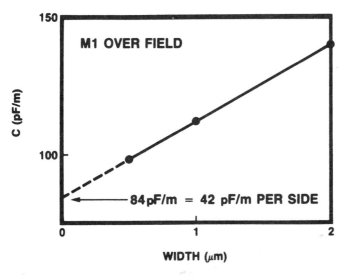

Fig. 15.2. Extraction of the fringing capacitance of a metal line over field oxide.

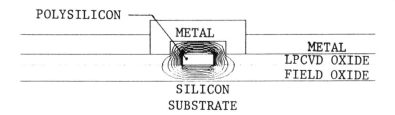

Fig. 15.3. Multi-layer capacitive coupling simulation.

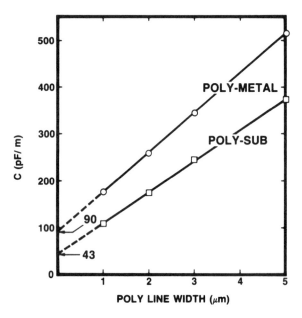

Fig. 15.4. Extraction of multi-layer fringing capacitances.

the capacitive coupling to the two layers. The major interest here is in the fringing components to the two layers. The areal capacitance between the polysilicon and the upper and lower layers can be calculated independently, but the fringing component will be "shared" between the two layers. Hence it is wrong to use the fringing capacitance coefficients of the polysilicon line to separate metal or substrate planes. The fringing component to the two layers in this structure can be simulated by using essentially the same method as described in the previous discussion of a conducting line over a conducting

METAL LINE OVER DIFFUSION

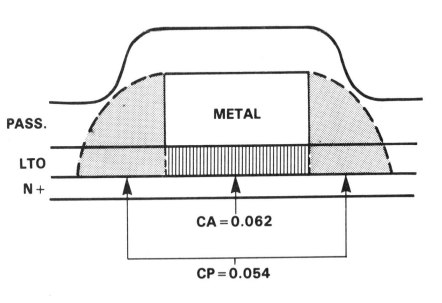

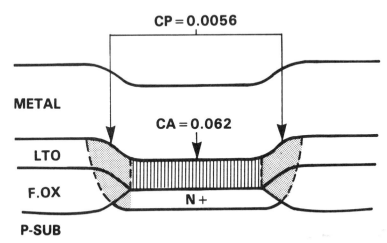

Fig. 15.5. Comparison of fringing capacitance between metal line to n^+ plane and n^+ line to metal plane; units are in $fF/\mu m^2$ and $fF/\mu m$ for area and peripheral components respectively.

substrate. By varying the polysilicon line width, the fringing capacitance to the metal and substrate can be calculated. Fig. 15.4 shows the extraction of the parameters. The simulation shows that when the polysilicon line is covered by metal on the top, the fringing capacitance of the polysilicon line to the substrate is reduced by 60 %, since part of the fringing field has terminated at the metal plane.

Another interesting situation is for the case of metal over diffusion. Fig. 15.5 shows the two cases of metal line over diffusion plane (upper figure) and metal plane over diffusion line. The measured parasitic capacitance (in units of fF/μm^2 for areal capacitance CA and fF/μm for fringing capacitance CP) are shown for both cases. The thickness of the metal, low temperature oxide (LTO) and field oxide (F.OX) are 1.0, 0.5 and 0.5 μm respectively. The fringing capacitance for the diffusion line to metal plane is much smaller than that of the metal line to diffusion plane. This is because at the side wall of the diffusion line, the effective dielectric thickness between the metal plane and the diffusion perimeter is much larger than the deposited oxide thickness. This is due to the shape of the field oxide in that region. This structure can be simulated by SCAP2 using the flexible input geometries described in section 4.2.

As mentioned in section 4.1, the interline capacitance of interconnections is becoming increasingly significant as the line spacings are reduced while the line thickness remains about the same. SCAP2 is used to simulate the case of metal lines with small spacings, and over a polysilicon plane. The simulated potential contour for the case of metal lines having different potentials is shown in Fig. 15.6. In this case, the line spacing is 0.8 μm and the line thickness is 0.6 μm. Fig. 15.7 shows the graph of the capacitance of the metal line versus line spacing for a case where the interlevel dielectric is a combination of nitride and oxide. The results show that the total line capacitance increases by 3E-3 fF/μm and 1.3E-2 fF/μm when the line spacings are 2 μm and 1 μm respectively. The increase in capacitance is also dependent on the distance of the lines from the underlying conducting plane. If the distance is reduced, the interline capacitance will be reduced, since more of the field lines will be terminated on the conducting plane, thus reducing the coupling between the metal lines. Fig. 15.8 compares the two cases where the metal lines are above the substrate and above the polysilicon plane. The simulated results show that

Examples of Parasitics Simulation

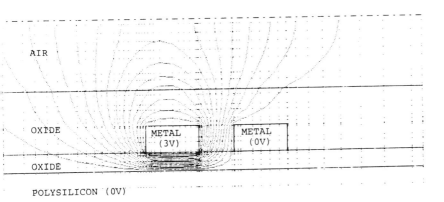

Fig. 15.6. Two-dimensional potential profile for two metal lines at different potentials above polysilicon plane.

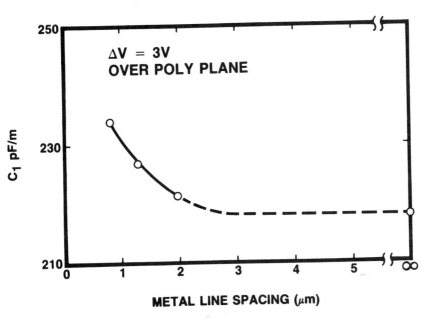

Fig. 15.7. Simulated total line capacitance per line vs. spacing for two metal lines at different potentials.

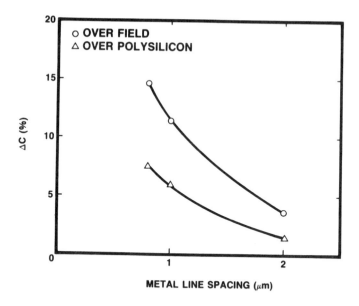

Fig. 15.8. Simulated increases in metal line capacitance vs. metal line spacing for two metal lines at different potentials.

the increase in the total metal line capacitance with reducing line spacing (ΔC) is larger for the former case.

In the case where the potentials of the metal lines are the same, the total capacitance of two lines to the substrate decreases as the lines get closer to each other. (The extreme case would be when the line spacing is zero, in which case the total fringing capacitance due to the side walls of the two lines would be reduced by a factor of two.) Fig. 15.9 shows the potential contour for this case. By using the same technique as was discussed for the case of lines with different potentials, the reduction in the parasitic capacitance is calculated, as shown in Fig. 15.10.

Field Capacitance Calculation by SUPREM and GEMINI

When the capacitance involves the formation of a depletion layer with a width which is a significant fraction of the dielectric thickness, the SCAP2 program is inaccurate. This occurs in the case of the field capacitor, where the substrate underneath the field oxide forms a depletion layer, the depth of

Examples of Parasitics Simulation

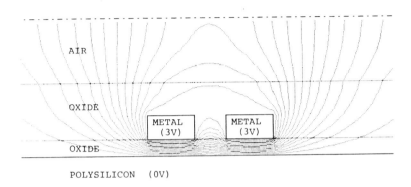

Fig. 15.9. Simulated two-dimensional potential for two metal lines at the same potential over polysilicon plane.

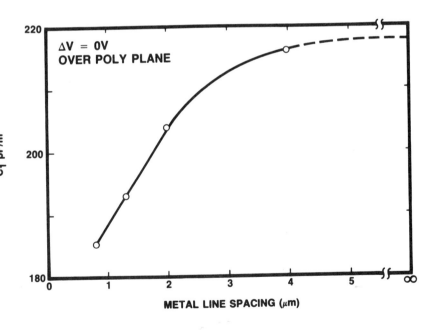

Fig. 15.10. Simulated total line capacitance per line vs. metal line spacing for two metal lines biased at the same potential over polysilicon plane.

which depends on the field implant profile. In this case, the depletion layer thickness is about 0.1 μm, while field oxide is typically 0.5 to 0.6 μm. Poisson's equation has to be solved for a given bias condition, and then the capacitance must be calculated. The SUPREM program is used to simulate the impurity profile of the field implant under the field oxide. Then the profile is transferred to the GEMINI program, which creates the field capacitor, with the desired bias on the capacitor electrode and substrate. Fig. 15.11 shows the

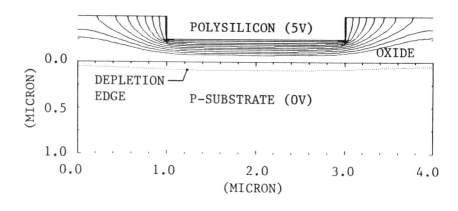

Fig. 15.11. Simulation of capacitance of polysilicon over field oxide by GEMINI.

potential contours from a GEMINI simulation of a polysilicon field capacitor. The symmetric boundary condition for the potential also allows one to consider the effect of periodic polysilicon lines biased at the same potential and separated by a specified spacing. This condition occurs in the polysilicon serpentine structure, and has the effect of reducing the fringing capacitance. Fig. 15.12 shows the extraction of the fringing and areal capacitances for an essentially isolated polysilicon line. The depletion layer formed at 5 V has significantly reduced the capacitance. Fig. 15.13 shows the effect of varying spacing between polysilicon lines at the same potential, with fixed line width. The reduction in the total capacitance is due to the reduction of the fringing component. The results shows that the reduction is not significant until the spacing is below 2μm.

Examples of Parasitics Simulation

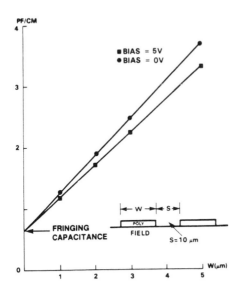

Fig. 15.12. Areal and fringing capacitance extraction of polysilicon line over field oxide.

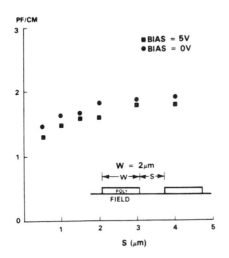

Fig. 15.13. Simulation of the total capacitance of polysilicon lines over field oxide vs. line spacing for the lines at the same potential.

Diffusion Capacitance Calculation by SUPRA

Diffusion capacitance simulation involves the simulation of the electric field in heavily doped regions. The capacitance is a function of the impurity profile in the structure. Fig. 15.14 shows the simulation by SUPRA of the

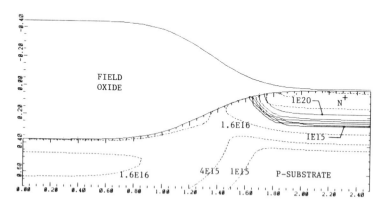

Fig. 15.14. Two-dimensional simulation of the n^+ profile by SUPRA. The dotted and solid contours are for boron and arsenic concentration (cm^3) respectively.

impurity profile of an n^+ diffusion region adjacent to the field oxide. The accuracy of the capacitance calculations depends very much on the accuracy of the doping profile. Careful analysis of the results shows that the accuracy of measurements is better than that of simulations. Thus measured data are used for circuit simulations. It is still useful to simulate the profile in this case because it provides a general picture of the impurity profile, and also indicates what will happen qualitatively to the capacitance if the processing steps are changed. The errors of the capacitance calculation reflect the difficulty of accurate process simulation for impurity profiles, particularly at the edges.

Interconnect Resistance Calculation by SCAP2

The resistance of a film of arbitrary two-dimensional geometry can be calculated using the SCAP2 program. Fig. 15.15 shows the simulation of the resistance of a two-dimensional shape. In order to use SCAP2 to calculate the

Examples of Parasitics Simulation

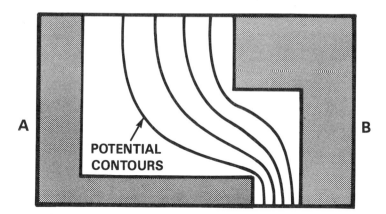

Fig. 15.15. Resistance calculation for arbitrary two-dimensional shape.

resistance from contact (shaded areas) A to B, the materials of the contacts are defined to be metallic, and set at different potentials. The capacitance between the two metallic contacts are then calculated by SCAP2. The capacitance is related to the surface integral of the electric field at the metallic surface (the contacts A and B). This result can be converted into current flowing between A and B by replacing the dielectric constant between the contacts used in the capacitance calculation by the conductivity of the material under study. The total current flowing through the two contacts can thus be found for the bias condition specified, hence giving the resistance between the contacts. In this example, the calculated effective sheet resistance is 0.38×(sheet resistivity). In other words, the 2-D shape between the two contacts is equivalent to 0.38 square in resistance.

Experiment and Simulation Comparisons

Measurements have been made on parasitic capacitance test structures with serpentine lines and are compared with simulated results. Examples of comparisons are shown in Table 15.1. The physical dimensions of the test structures were measured by SEM, and used in the simulations. Results show that the agreement is, in general, quite good. Note that in several cases of multi-layer structures such as polysilicon lines over field oxide and under a

	C (area) (fF/μm^2)	C(periphery) (fF/μm)
Polysilicon line over field		
Measurement	0.053	0.046
Simulation	0.056	0.035
Metal line over diffusion		
Measurement	0.062	0.061
Simulation	0.069	0.060
Metal line over polysilicon		
Measurement	0.067	0.061
Simulation	0.069	0.058

Table 15.1. Measured and simulated data of parasitic capacitances.

metal plane, a combination of simulations and measurements is necessary to obtain the coefficients. Much improvement in the accuracy of circuit simulations has been obtained by using this methodology of parasitic capacitance extraction.

15.3 Three-Dimensional Parasitic Components Extraction

Muti-level Interconnect Capacitance Calculations by FCAP3

A three-dimensional simulation is often necessary for the interline capacitance calculation in a multi-level interconnect system. To demonstrate the importance of a three-dimensional configurations for this problem, three situations of two-level metal lines over a ground plane are simulated: parallel lines, 45° crossing lines, and 90° crossing lines as shown in Fig. 15.16(a), (b), and (c). The interline capacitance values are calculated with respect to the vertical spacing between the first and second level metal by three different methods. The simplest method is a one-dimensional calculation. It does not consider any fringing field, that is, the capacitance values are obtained by multiplying the overlapping area with the areal coefficient for an infinite planar capacitor. The second method is a two-dimensional calculation. It includes the two-dimensional fringing field using SCAP2, but it still does not include the three-dimensional fringing at the corners of the crossing lines. The third and

Examples of Parasitics Simulation

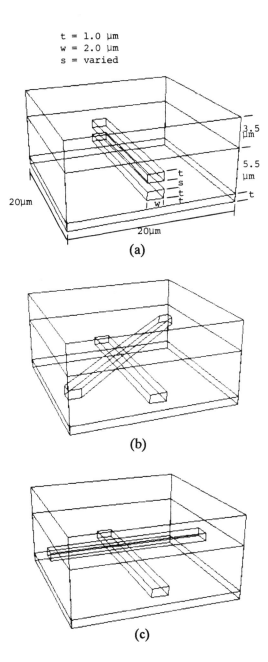

Fig. 15.16. Three cases of two-level on-chip metal configurations. (a) parallel (b) 45° crossing (c) 90° crossing.

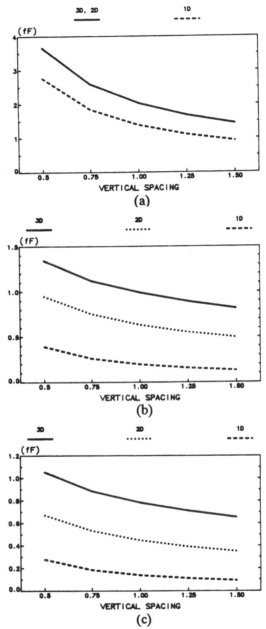

Fig. 15.17. Capacitance calculation for the three cases in Fig. 15.16 by FCAP3.

Examples of Parasitics Simulation 351

complete calculation is done three-dimensionally by FCAP3.

Fig. 15.17(a) shows the interline capacitance for the parallel lines in Fig. 15.16(a). Since the lines are parallel, the two-dimensional calculation is an accurate simulation and is the same as the three-dimensional calculation. As the vertical spacing becomes larger, the capacitance decreases for both the two-dimensional and one-dimensional calculations. However, the relative error of one-dimensional calculation to the two-dimensional calculation becomes larger since the effect of the fringing field increases. For the 45° crossing lines in Fig. 15.16(b), we have to do the three-dimensional simulation to obtain a correct capacitance. As shown in Fig. 15.17(b), the two-dimensional calculation underestimates the capacitance given by the three-dimensional calculation, and the one-dimensional calculation underestimates it even more. Since the 90° crossing has more three-dimensional characteristics than the 45° crossing, two-dimensional and one-dimensional calculations become almost useless. For the spacing of 1.5 μm, the result in Fig. 15.17(c) shows that the two-dimensional calculation estimates the capacitance to be only half of the correct capacitance calculated by the three-dimensional calculation.

The above examples emphasize the importance of the three-dimensional simulation which most current circuit extractors do not consider. When they extract a circuit from its layout including interconnect capacitances, the extractors usually calculate them using a two-dimensional approximation, namely, areal and peripheral coefficients. But the smaller the size of interconnects and the greater the number of interconnect levels, the more inaccurate this approach is.

Fig. 15.18 shows another example of capacitance calculation of two-level interconnect lines. This time, three different dielectric materials are used to be more realistic. Due to the symmetry, half of the whole problem is simulated by cutting with a xz-plane along the center of the second level metal (note the triad in the figure for the directions). Table 15.2 lists the amounts of charge on each conductor with a bias condition. Also, the potential plots in three ways - bird's-eye-view plot, contour plot, and one-dimensional plot - are shown in Fig. 15.19(a), (b), and (c). The bird's-eye-view and contour plots are the potential distribution on the plane, $z = 2.4$ μm, and the one-dimensional plot is

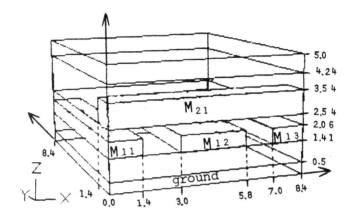

Fig. 15.18. Two-level on-chip metal lines simulated by FCAP3.

Conductor	Potential	Charge (fC)
ground	0 V	-1.469
C_{11}	0 V	-0.338
C_{12}	1 V	3.023
C_{13}	0 V	-0.448
C_{21}	0 V	-0.768

Table 15.2. Capacitance values of the structure in Fig. 15.18 calculated by FCAP3.

the potential plot along the line, $x = 4.1\ \mu m$ and $z = 2.4\ \mu m$. The plots show the high potential along the first-level center conductor (M12) and the suppressed potential due to the second-level conductor (M21) near the xz-plane.

It is difficult to benchmark the above calculations against real measurements, since the values are too small to be measured directly. In fact, this is one of the main reasons that we have to resort to simulations to estimate the values. To verify the accuracy of three-dimensional calculations by FCAP3, we can build a scaled up structure and measure the capacitance of it. Fig. 15.20 shows a structure with metal bars and Table 15.3 lists the measured capacitance values for different configurations. The capacitance values simulated by FCAP3 and percentage difference between the measured and simulated values

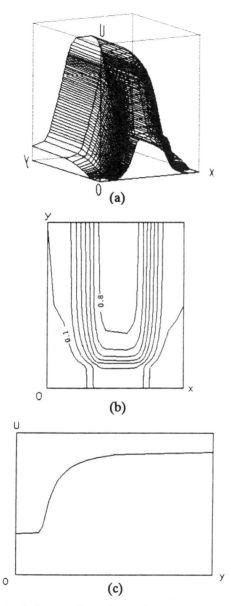

Fig. 15.19. (a) Bird's-eye-view plot and (b) contour plot of potential at the plane $z=2.4\mu m$ of the structure in Fig. 15.18 calculated by FCAP3. (c) Potential plot along the line $x=4.1\mu m$ $z=2.4\mu m$.

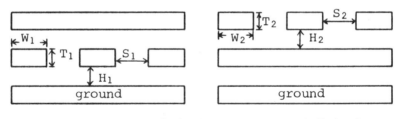

$W_1 = W_2 = 1$ inch, $T_1 = T_2 = 0.5$ inch

Fig. 15.20. Side views of the structure used to benchmark the FCAP3 calculations.

| H_1 | S_1 | H_2 | S_2 | | C_0 | C_{1g} | C_{11} | C_{12} | C_0-C_{1g} |
	(inches)					(pF/11inches)			-C_{11}-C_{12}
0.30	1.00	1.00	1.00	Measured	19.94	13.13	2.50	4.06	0.25
				FCAP3	19.40	12.97	2.35	4.08	0.00
				Diff(%)	-2.71	-1.22	-6.00	0.49	
1.03	1.00	0.25	1.00	Measured	18.81	4.21	2.88	11.38	0.34
				FCAP3	18.88	4.25	2.44	12.19	0.00
				Diff(%)	0.37	0.95	-15.28	7.12	
0.37	1.00	1.00	4.50	Measured	17.77	11.66	3.17	2.29	0.65
				FCAP3	16.84	11.02	3.25	2.57	0.00
				Diff(%)	-5.23	-5.49	2.52	12.23	
1.27	5.00	0.60	4.50	Measured	10.74	5.78	0.17	4.39	0.40
				FCAP3	10.51	5.28	0.22	5.01	0.00
				Diff(%)	-2.14	-8.65	29.41	14.12	
1.27	5.00	0.60	10.00	Measured	9.58	6.38	0.44	2.45	0.31
				FCAP3	9.35	5.87	0.60	2.88	0.00
				Diff(%)	-2.40	-7.99	36.36	17.55	
1.27	10.00	0.60	10.00	Measured	9.44	6.64	0.09	2.50	0.21
				FCAP3	9.35	6.18	0.17	3.00	0.00
				Diff(%)	-0.95	-6.93	88.89	20.00	

Table 15.3. Comparison of measured and simulated capacitances of the structure in Fig. 15.20.

are also listed in Table 15.3. The conservation of charge requires $C_{11}+C_{12}+C_{1g}-C_0$ to be zero, where C_{11}, C_{12}, C_{1g}, and C_0 are the capacitances between the first-level center metal and the two adjacent first-level metals, the capacitance between the first-level center metal and the three second-level metals, the capacitance between the first-level center metal and the ground, and the total capacitance of the first-level center metal, respectively. As shown in Table 15.3, this is consistently satisfied by all the values from FCAP3. On

Examples of Parasitics Simulation

the contrary, the values from the experiment have some offsets. We can interpret the offsets are due to a measurement error. This measurement error causes large percentage difference between the simulation and measurement for the small capacitance values. However, if the measurement error is considered, Table 15.3 shows an excellent agreement between the measurement and the simulation by FCAP3.

Via Hole Capacitance Calculation by FCAP3

The capacitance between a conductor plane and a line which goes through a hole in the plane is important to calculate accurately in modeling of the integrated circuit packages and printed circuit boards. We can easily simulate the structure using the second level input scheme of FCAP3. We need only simulate one-eighth of the structure, taking advantage of the symmetry of the structure as shown in Fig. 15.21(a). Fig. 15.21(b) shows the calculated capacitance values with respect to the line diameter. The two-dimensional approximation is obtained by multiplying the thickness of the plane and the

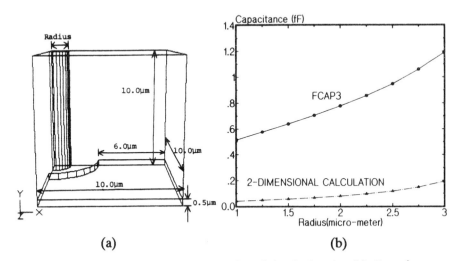

Fig. 15.21. (a) A via hole and a line through the via. (b) Capacitance calculations of the via structure v.s. the radius of the line by FCAP3 and two-dimensional approximation.

capacitance per unit length of infinitely long cylinders. It underestimates the capacitance, especially when the diameter of the line is small, because it does not include the fringing field.

Via Resistance Calculation by FCAP3

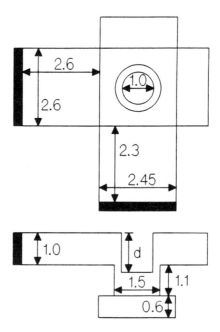

Fig. 15.22. A via connection of two-level on-chip metal lines. The shaded areas are contacts to the metal layers for the simulation

d (μm)	Resistance (Ω)
0.0 (plug)	0.1237
1.1	0.1292
1.35	0.1306
1.6	0.1320
1.85	0.1329

Table 15.4. Resistance of the via connection in Fig. 15.22 by FCAP3.

FCAP3 can calculate the resistance of a three-dimensional structure using the duality between capacitance and resistance as described in section 4.3. With this method, the resistance of a via connect in Fig. 15.22 is calculated varying the depth of the void at the center. The shape of the void depends on the multi-level metal process used. The result in Table 15.4 shows that the resistance does not change significantly. This is reasonable because the void is located where the current density would be low even if the void were filled [15.3].

References

[15.1] S. Seki and H. Hasegawa, "Analysis of Crosstalk in Very High-Speed LSI/VLSI's Using a Coupled Multiconductor MIS Microstrip Line Model," *IEEE Trans. Electron Devices,* **ED-31,** no. 12, pp.1948-1953, Dec. 1984.

[15.2] H. Hasegawa and S. Seki, "Analysis of Interconnection Delay on Very High-Speed LSI/VLSI Chips Using an MIS Microstrip Line Model," *IEEE Trans. Electron Devices,* **ED-31,** no. 12, pp.1954-1960, Dec. 1984.

[15.3] K. Lee, P. Vande Voorde, M. Varon, and Y. Nishi, "A Method to Reduce the Peak Current Density in a Via," *Tech. Digest of IEDM 1987.*

Appendix

Source Information of 2-D programs

SOURCE INFORMATION OF 2-D PROGRAMS

PROGRAM	SOURCE	ADDRESS	
SUPREM	Stanford Univ.	Office of tech. licensing, Stanford Univ. 105 Encina Hall, Stanford, CA 94305	Public
SUPRA	"	"	"
SOAP	"	"	"
GEMINI	"	"	"
CADDET	Hitachi	Tech. Administration, Hitachi P.O. Box 2, Kokubunji, Tokyo, Japan	Technical exchange
PISCES	Stanford Univ.	Office of Tech. licensing, StanfordUniv. 105 Encina Hall, Stanford, CA 94305	Public
FCAP2	Hewlett Packard	Hewlett Packrad Lab. 3500 Deer Creek Rd., Palo Alto, CA 94304	Technical exchange
TECAP2	"	"	Commercial
HPSPICE	"	"	Commercial

Table of Symbols

A_{ij}	sensitivity matrix element
A	angstrom unit (1E-10 m)
B	parabolic oxide growth-rate constant
B	empirical parameter in mobility model
B/A	linear oxide growth-rate constant
BV_{DS}	source-drain breakdown voltage
C	capacitance
C	impurity concentration
C_o	oxidant concentration at the oxide interface
C_o	capacitance of a transmission line with dielectric materials removed
C_o	output loading capacitance
C_1, C_2	doping concentrations
C_A	areal capacitance coefficient
C_D	depletion capacitance
C_D	drain junction capacitance
C_i	oxidant concentration at the silicon interface
C_n	coefficient of auger recombination
C_n	n^- pocket characteristic length
C_{ox}	gate oxide capacitance

C_P	peripheral capacitance coefficient
C_p	coefficient of auger recombination
C_p	peak concentration
C_T	total impurity concentration
C_v	normalized vacancy density
$C^\dagger$	electrically active impurity concentration
C^*	equilibrium oxidant concentration in the oxide
c	speed of light in vacuum
D	diffusivity
D_C	counter-doping dose
D_{eff}	effective diffusion coefficient
D_i^+	diffusivity due to positive vacancies
D_i^-	diffusivity due to negative vacancies
$D_i^=$	diffusivity due to doubly negative vacancies
D_i^x	diffusivity due to neutral vacancies
d_k	mask thickness at k^{th} segment
D_n	n^- pocket dose
D_N	diffusivity under non-oxidizing condition
D_{ox}	diffusivity due to oxidizing ambient
E	electric field strength
E_c	electron energy of conduction band edge
E_{fp}	quasi-Fermi level for holes
E_i	electron energy at the intrinsic Fermi level
E_m	maximum electric field
E_n	electric field normal to current flow
E_p	electric field parallel to current flow
E_{SAT}	critical field for velocity saturation
E_μ	activation energy of oxide viscosity
F, fF, pF	Farad unit, 1E-15 F, 1E-12F
F, F_1, F_2, F_3	oxidants fluxes

Table of Symbols

G	empirical parameter in mobility model
G_{DS}	channel conductance
g_m	transconductance
g_z	oxide growth rate in the z-direction
h	transport coefficient
I	current
$I(x,y)$	Implant profile
I_{DD}	maximum available current from power supply
I_{DS}	drain to source current
I_{DSAT}	drain to source current at saturation
I_H	latchup holding current
I_L	transistor subthreshold leakage current
I_{max}	maximum concentration
I_{PT}	punchthrough current
I_{SUB}	substrate current
I_T	latchup triggering current
I_{TH}	transistor threshold current
J	impurity flux
$\mathbf{J_n}$	electron current density
$\mathbf{J_p}$	hole current density
J_o	coefficient of current density
K	scaling factor
k	Boltzmann constant
k	surface reaction coefficient
L	inductance
L	channel length
L	decay length in thin oxide regime
L_D	length of uniform drain region
L_{eff}	effective channel length

L_{min}	minimum channel length
L_p	polysilicon gate length
L_S	length of uniform source region
L_v	diffusion length of vacancies
L^*	effective length of potential barrier
m	segregation coefficient
N	N_A-N_D
N	number of nodes
N^+	doping concentration in n$^+$ polysilicon
N_A	acceptor concentration
N_A^-	ionized acceptor concentration
N_B	substrate doping
N_c	correction concentration for oxidation
N_D	donor concentration
N_D^+	ionized donor concentration
N_i	intrinsic carrier density
N_i	dopant concentration, inert ambient
N_{ix}	concentration in x-direction, inert ambient
N_{iy}	concentration in y-direction, inert ambient
N_o	coefficient of band gap narrowing
N_o	dopant concentration, oxidizing ambient
N_{ox}	concentration in x-direction, oxidizing ambient
N_{oy}	concentration in y-direction, oxidizing ambient
N_p, N_p'	implantation dose
N_S	substrate surface impurity concentration
N_w	N-well doping concentration
n_s	surface electron concentration
n	unit vector normal to the surface
n	electron density
n_i	intrinsic carrier density
$n_{i,e}$	effective intrinsic carrier density
nm	nanometer unit (1E-9 m)

Table of Symbols

P	Gaussian probability distribution
P	pressure
p	hole density
Q	minority carrier concentration in GEMINI
Q_B	integrated carrier in GEMINI punchthrough simulation
Q_L	integrated carrier in GEMINI channel-length simulation
Q_{NA}	channel implant dose
Q_{ND}	depletion implant dose
Q_n	electron density per unit area in the channel
Q_{ss}	interface fixed charge density
Q_W	integrated carrier in GEMINI channel-width simulation
q	electron charge
R	resistance
R	recombination
R_D	drain series resistance
R_n	N^- pocket peak depth
R_p, R_p'	range of Gaussian implant distribution
R_S	source series resistance
S	subthreshold slope
S	empirical parameter in mobility model
S_p	poly gate side-wall spacer thickness
T	absolute temperature
T_{ox}	oxide thickness
t	time
U	active impurity concentration
$U(n,p)$	electron and hole recombination rate

V	oxide growth velocity
V^-	negative vacancies
$V^=$	doubly negative vacancies
V^+	positive vacancies
V_B	substrate bias voltage
V_{BS}	substrate to source bias voltage
V_C	empirical parameter in mobility model
V_D	drain bias voltage
V_{DD}	circuit bias voltage
V_{DS}	drain to source bias voltage
V_{DSAT}	saturation drain voltage
V_{fb}	flat-band voltage
V_G	gate bias voltage
V_{GS}	gate to source bias voltage
V_i	i^{th} device parameter in sensitivity matrix
$\overline{V_i}$	nominal value of V_i
V_{in}	inverter input voltage
V_{out}	inverter output voltage
V_{PT}	punchthrough voltage
V_p	barrier potential
V_S	source bias voltage
V_s	saturation velocity
V_{SB}	source bias with respect to the substrate
V_{SS}	circuit ground voltage
V_T	transistor threshold voltage
V_{T0}	threshold voltage at zero substrate bias
$V_{T(LOW)}$	low drain bias threshold voltage
$V_{T(HI)}$	high drain bias threshold voltage
V_{TLC}	long channel threshold voltage
V_{TN}	n-channel threshold voltage
V_{TP}	p-channel threshold voltage
W	transistor channel width
W_D	drawn channel width

Table of Symbols

W_{eff}	effective channel width
x	x-coordinate
X	process feature
$\overline{X}$	average value of process feature
X_d	depletion depth
X_i	initial oxide thickness
X_j	diffusion junction depth
X_0	oxide thickness
Y_j	counter-doping junction depth
y	y-coordinate
Z_c	characteristic impedance
Z^*	effective width of potential barrier
ΔL	channel length variation
ΔR_p	standard deviation in y-direction
ΔW	channel width loss per side
Ω	material boundaries
Θ	vector stream function
∇	gradient operator
α	empirical parameter in mobility model
$\delta\phi_n$	increment of electron quasi-Fermi potential
$\delta\phi_p$	increment of hole quasi-Fermi potential
ε	dielectric constant
μ	mobility
μ	oxide viscosity
μ_i	ith moment of an implant distribution
μ_n	electron mobility
μ_o	zero-field mobility
μ_p	hole mobility
ϕ	electric potential

ϕ_p	hole quasi-Fermi potential
ϕ_n	electron quasi-Fermi potential
ψ	internal potential
ψ_B	difference between intrinsic and hole Fermi level
ρ	oxide density
ρ	resistivity
σ	standard deviation
σ_p, σ_p'	standard deviation of Gaussian implant distribution
τ_n	electron life time
τ_p	hole life time
θ	z-component of stream function
1-D	one-dimensional
2-D	two-dimensional

Subject Index

Avalanche breakdown, 171, 212, 295

BIRD, 14
Bird's beak, 235, 259
Body effect, 188-190
Boron encroachment
   LOCOS, 252, 258
   modified LOCOS, 259-261
   SWAMI, 265-267
Boundary-value method, 61
Breakdown, source-drain, 171, 212, 295
Buried channel, 328

CAD, 8
CADDET
   basic equation, 90-92
   device simulations, 97-101, 211-230, 315-333
   flow chart, 92-94
   grid, 92-93
   input file, 98
   input format, 97-100
   input structure, 88-89
   mobility model, 93-97
   numerical technique, 92
   output, 101
CMOS, 271-273
   cross-section, 25
   n-well, 235, 277
   p-well, 246, 248
   submicron, 271-273
   trench isolated, 233
Capacitance
   depletion layer, 170
   gate oxide, 170
Channel implant
   deep, 187
   depletion, 328-333
   n-channel, 186-188

shallow, 187
Channel length, effective, 183, 317
Channel profile
   n-channel, 179, 182
   p-channel, 279, 290
Channel width
   drawn, 252
   effective, 167, 183, 252, 257
   extraction of effective, 255-257
   loss, 255-257, 268
Channeling, 279
Characteristic impedance, 133
Counter-doping, 144, 277-280
   dose, 144, 278
   energy, 144
   junction, 279-280
Critical electric field, 212
Current continuity equation
   stream function form, 91
   time dependent form, 103
Current transport equation, 104

Degradation
   hot electron, 212-213, 295
   linear transconductance, 217
   saturation transconductance, 217
Depletion layer capacitance, 170
Depletion mode MOSFET, 326-333
Device optimization, 174
Device simulation, history, 13-14
DIBL (see Drain-induced barrier lowering)
Diffusion
   across interfaces, 38-39

   concentration enhanced, 163
   continuity equation, 35
   field-driven, 35-36
   gradient-driven, 35
   high concentration, 36-38, 51-53, 163
   low concentration, 37, 48-49
   moving boundary, 49-51
   oxygen enhanced, 38, 157-162
   vacancy assisted, 36-37
Diffusivity
   arsenic, 37
   boron, 37, 157
   effective, 59
   intrinsic, 36-37
   phosphorus, 37-38
Direct matrix solution, 106
Double diffused drain, 213, 295
Drain electric field, 221-229
Drain-induced barrier lowering, 197-207, 280-289
Duality, 135

Einstein's relation, 90
Electric field
   channel, 212
   critical, velocity saturation, 212
   drain, 212, 221-229
   maximum, drain, 212, 225-226
   normal, 94
   parallel, 94
Electromigration, 4

FCAP2, 130

Subject Index

FCAP3, 136
FIELDAID, 14
Factorial experiments, 144
Fermi-Dirac integral, 75
Fermi-Dirac statistics, 90
Finite-difference method, 79
Flat-band voltage, 326
Fringing component, 338

GEMINI
  basic equation, 75-77
  device simulations, 177, 181-184
  capability, 72-73
  channel length simulation, 80-83
  channel width simulation, 83-85
  five-point finite-difference approximation, 79
  grid, 77
  input file, 80-86
  numerical technique, 79
  punchthrough current calculation, 76-77
  SWAMI simulations, 265-268
  trench simulations, 238-245
Gate oxide
  capacitance, 180
  effect on subthreshold slope, 284-285
Gaussian implant profile, 27
Glass-transition temperature, 59
Graphical post-processing, 16
Green's function, 14-15, 61
Gummel's iteration, 92, 105
Gummel-Scharfetter formula, 105

Hot electron degradation, 174, 212, 295

ICCG, 105, 130
Impact ionization, 212
Implantation energy, 324
Impurity profile
  n^- pocket, 303
  n-channel, 45, 179, 182
  p-channel, 158, 279, 290
  source/drain, 45, 68, 179
  SWAMI, 266
  trench structure, 239
Inductance, 133
Interconnect capacitance, 335-345, 348-355
Interconnect resistance, 346-347
Interface damage, 212
Inverter circuit, 181, 320
Ion implantation
  boron, 153-154
  distribution moments, 29-30
  gaussian profile, 27
  Pearson IV profile, 30, 153-154
  projected range, 29
  standard deviation, 29
Isolation, 251-269

Jacobian matrix, 106

LDD (see lightly-doped drain)
LOCOS, 252-258
  modified, 259-261
Latchup, 233-234

holding current, 246-247
initiating current, 233, 246-247
path, 234
Lightly-doped drain, 174, 211-230, 297-298
  processing, 214-215
  simulation, 216, 297-298
Linear transconductance, 217
Linear-parabolic oxide growth model, 32

MINIMOS, 14
MOSFET, 1-2
  depletion mode, 326-333
  enhancement mode, 317-326
Maxwell-Boltzman statistics, 90
Minimum overlap device, 229
Mobility
  electron, 171, 317
  electron mobility model in CADDET, 97
  Gummel's bulk mobility model, 94
  hole mobility model in PISCES-II, 104
  modified Gummel's mobility model, 97
  normal field mobility model for electron, 97
  reduction, 320
Moving-boundary problem, 18

N-well
  CMOS, 233-234, 277
  dose, 146

Narrow width effect, 83-85, 235, 251
Navier-Stoke's equation, 60-61
Newton method, 105-106, 116
Newton-Richardson procedure, 106, 116
Numerical simulation system
  block diagram, 16-18
  hierarchical simulation, 15
  implementation scheme, 16-18
  system support, 15-16

Obtuse triangle problem, 107
Oxidation
  linear-parabolic growth model, 32
  non-planar, 58-61
  thin oxide, 35, 154-156
  two-dimensional model, 59-61
  volume expansion, 31
Oxide
  incompressibility, 61
  thinning, 58
  viscosity, 59
Oxide isolation SUPRA simulation, 53-54
Oxide encroachment
  LOCOS, 252-255
  SWAMI, 263

PISCES-II
  1-D plot, 123-125
  2-D contour plot, 122-123
  I-V plot, 120-121
  basic equation, 103
  electrical characteristics

Subject Index

simulation, 113-119
  grid generation, 107-111
  initial guess, 106
  input file, 108
  numerical algorithm, 105
  output, 119
PLPKG (see Plotpackage)
Packing density, 235, 272
Parasitic bipolar transistors, 233-234
Parasitic components, 129
Pearson IV profile, 30, 153-154
Performance, transistor, 174, 216-217
Plot package(PLPKG), 16-17, 72-73, 88
Poisson equation
  in CADDET, 90
  in GEMINI, 75
  in PISCES-II, 103
Polysilicon gate
  n^+, 275-276
  p^+, 275
Potential barrier height, 201
Potential profile
  n^- pocket, 305
  n-channel, 182, 295-296
  p-channel, 280-283
  trench structure, 240-244
Power supply voltage, 181, 315
Process development, 6-8
Process simulation:history, 14-15
Program interface, 16
Projected range, 29
Punchthrough, 185-186, 198
  bulk, 201-206
  current, 185
  n-channel, 324, 197-208
  p-channel, 206-207, 281
  point, 203
  surface, 201-206
  voltage, 185, 198, 324

Quasi-Fermi level, 183, 240

Recombination
  Auger, 104
  Shockley-Reed-Hall, 104
Reliability, MOSFET, 212, 298
Residual current
  n-channel, 169
  p-channel, 285-286

SCAP2, 130
SIFCOD, 14
SLOR (see Successive line over-relaxation)
SOAP
  boundary conditions, 59-60
  flow chart, 62
  LOCOS simulation, 63-66
  oxidation models, 59-61
  SWAMI simulation, 263-264
SPICE, 19
SUPRA
  device simulations, 178, 253, 265, 297
  diffusion models, 49-53
  ion implantation models, 47-48
  LDD structure, 297

LOCOS simulation, 253-256
NMOS transistor simulation, 53-57
SWAMI simulation, 265-266
SUPREM
  boron implantation, 153-154
  diffusion models, 35-39
  ion implantation models, 27-31
  n-channel threshold simulation, 278
  NMOS transistor simulation, 39-45
  oxidation models, 31-35
  oxygen enhanced diffusion of boron, 157-162
  p-channel threshold simulation, 144-148, 181, 278
  thin oxide growth, 154-156
SWAMI, 261-268
Saddle point, 203
Saturation region, 171
Saturation transconductance, 217
Saturation velocity, 171, 273
Scaling, 1-6
  constant field, 1-6
  constant voltage, 3-4
  depletion mode, 326-333
  enhancement mode, 315-326
  factor, 4-6
  history, 1-2
Segregation coefficient, 39, 157
Series resistance, 217, 317
Short channel effects, 3, 181, 190, 322-324
Simulation
  basic device parameters, 180-186

  methodologies, 143-148
  tools, 175-180
Standard deviation, 29
Stone's method, 93
Stream function, 91-92
Stress, 263-264
Stress-relief oxide, 253
Strip line, 133
Substrate current, 212-213, 224, 227
Subthreshold characteristics, 168-169, 184
Subthreshold leakage, 169, 286
Subthreshold slope, 169-170, 281
  n-channel, 169-170
  p-channel, 281, 284-285
Successive line over-relaxation(SLOR), 79
Surface reaction rate coefficient, 31-32

TECAP2, 19
Threshold current, 83, 183
Threshold voltage, 83, 160, 168, 170, 183, 276
  analytical calculation, 276
  depletion mode MOSFET, 329
  simulation, 180, 183
  vs. channel length, 288, 190, 297
  vs. channel width, 258
Transconductance, 273
  linear, 217
  saturation, 217, 273
Transistor
  performance, 174, 211

Subject Index

current drive, 174, 273
Transport coefficient, 31-32, 39
Trench, 233
   CMOS, 233-235
   surface inversion, 235-248

Velocity Saturation, 4
Via hole capacitance, 355
Via hole resistance, 356
Viscosity, 59
Voltage standard, 4

About the Authors

Kit M. Cham was born in Hong Kong. He received the Ph.D. degree in applied physics from Yale University, New Haven, CT, in 1980. Since then, he has been with the Integrated Circuit Laboratory of Hewlett-Packard, Palo Alto, CA. He has worked on the characterization of MNOS devices for nonvolatile memories from 1980 to 1981. Since then he has been involved in the development of submicron CMOS technology. He is currently project leader for submicron CMOS device design. His work has resulted in 30 papers and two patents pending. He is coauthor of a book titled *Computer-Aided Design and VLSI Device Development* published by Kluwer Academic Publishers in 1986.

Soo-Young Oh received his Ph.D. in Electrical Engineering from Stanford University in 1980. He then joined Hewlett-Packard, and is now project manager of device, process, and system package modeling group in Structure Research Laboratory. He is currently working on a complete numerical simulation CAD system from process/device to circuit/system package. His work has resulted in 17 publications. He is a coauthor of a book titled *Computer-Aided Design and VLSI Device Development* published by Kluwer Academic Publishers in 1986.

John L. Moll received his Ph.D. in Electrical Engineering from Ohio State University in 1952. He is currently senior scientist and manager of the Integrated Circuit Structure Research Laboratory of Hewlett-Packard. His extensive work in solid state devices has resulted in over 100 papers and ten patents. He is author of a book, *Physics of Semiconductors*, and coauthor of a book titled *Computer-Aided Design and VLSI Device Development*. He is a fellow of IEEE, member of the American Physical Society, National Academy of Engineering, National Academy of Sciences, and Sigma Xi.

Keunmyung Lee was born in Korea in 1957. He received his M.S. and Ph.D. degrees from University of California, Berkeley, in 1982 and 1985, respectively, both in electrical engineering and computer sciences. Since 1985 he has been with Hewlett-Packard Laboratories, Palo Alto, CA. His current research interests are design and modeling of interconnect systems of integrated circuits and packages. He has published 8 papers, and has one patent pending.

Paul Vande Voorde recieved his Ph.D. degree in solid state physics from the University of Colorado in 1980. His dissertation topic involved fluctuation phenomenon in III-V semiconductors. He has worked at Hewlett-Packard Laboratories since 1981. While at HP Laboratories, Paul has been involved in several silicon processing projects related to advanced CMOS. Currently he is working on process and device simulation to support CMOS and Bipolar process development. He has authored or coauthered 13 publications in journals and conference proceedings.

Daeje Chin was born in Korea in 1952. He received the Ph.D. in electrical engineering from Stanford University in 1983. He developed the two-dimensional process simulation programs SUPRA and SOAP, for his dissertation at Stanford while working part-time at Hewlett-Packard I.C. Laboratory. After obtaining his doctorate, he joined IBM Watson Research Center in New

York. He has been working on submicron memory processing technology and high performance DRAM circuit design. In 1986, he joined Samsung Semiconductor, and is currently working on the development of 4MB DRAM. He has 15 publications and four patents pending. He is also coauthor of a book titled *Computer-Aided Design and VLSI Device Development* published by Kluwer Academic Publishers in 1986.